AF564800

MICROBIAL GENETICS

MICROBIAL GENETICS

By

DR. S. SUNDARA RAJAN

Principal

Reva Institute for Science and Technology Studies

R.T. Nagar

Bangalore – 560 024

ANMOL PUBLICATIONS PVT. LTD.

NEW DELHI - 110 002 (INDIA)

ANMOL PUBLICATIONS PVT. LTD.
4374/4B, Ansari Road, Daryaganj
New Delhi - 110 002
Ph.: 3261597, 3278000
Visit us at: www.anmolbooks.com

Microbial Genetics

First Edition, 2003

ISBN 81-261-1335-9

PRINTED IN INDIA

Published by J.L. Kumar for Anmol Publications Pvt. Ltd., New Delhi - 110 002 and Printed at Arora Offset Press, Delhi.

CONTENTS

PREFACE

Hereditary potentialities of living beings are same both for microorganisms as well as higher organisms. The mechanism of genetic transfer having originated in microorganisms first has reached a much more advanced stage in higher organisms.

Microbial Genetics traces the various patterns of genetic transfer and recombination in microorganisms in all their manifestations. The book consists of 11 chapters encompassing all aspects of microbial genetics such as structure of nucleic acids, their replication, mutation, recombination etc. There is also a chapter on genetic engineering.

The book is meant to be a text book for students of life sciences, pharmacy, agriculture etc. at the graduate and post graduate level.

I am thankful to Mr. J.L. Kumar, M.D. of Anmol Publications Pvt. Ltd., New Delhi for publishing this book.

—Dr. S. Sundara Rajan
Principal
Reva Institute for Science and Technology Studies
R.T. Nagar
Bangalore – 560 024

INTRODUCTION

"Like begets like" and "Chip of the old block" are two of the adages quoted often whenever we come across a situation to compare parents an children. These popular adages are indeed so commonplace that we may not be able to appreciate the complete truth behind it. Indeed why should a human being give birth to a human being and not to any other organism. A mango seed sown in soil produces only a mango tree and not a guava tree. What is it that is hidden in the mango seed which makes it grow into another mango tree only. When children are born we often hear statements like "the nose is like the father, the eyes are like the mother, but the mouth and hair are different". This shows that the new born in addition to having parental characters has its own haracters too. again what is the mystery behind the fact, that the child has sufficient similarity with the parents and yet sufficiently different from the also?

Genetics is a branch of biology that aims to find answers to the above questions. In fact, genetics is still trying to unravel the total mystery behind the statement, like begets like, even though most of the facts both at the morphological and at the molecular level have been understood.

The factg that children resemble their parents is due to heredity and they also differe from their parents to some extent is because of Variation. Heredity may be defined as the potential of an individual to transmit its characters to the offspring while variations are the differences that the individual acquires due to the interaction with the environment. The main aim of Genetics is the study of these seemingly opposite phenomena namely heredity and variation. While heredity tries to maintain uniformity, variation brings in plasticity.

The term Genetics was first used by Bateson in 1906, even though the science of genetics made its firm beginnings in the form of Mendel's epoch making work with pea plants. The rediscovery of Mendel's work independently by

Devries. Tschermark and Correns indeed paved the way for founding the science of Genetics even though the name was coined only in 1906.

Organisms are capable of transforming their characters to the next progeny as a result of reproduction. Indeed ever since life first arose on earth (due to a chance combination of elcments), it has maintained itself till to day in myriad forms only because of its capacity of reproduction. Untill 1868 when Mendel working with pea plants came out with coinvincing explanation regarding the mechanisms of heredity, it was largely speculative as to how organisms pass on their characters to the progeny.

Earlier concepts of heredity

Earlier concepts of heredity made their beginning first with the discovery of sexualityin organisms-both plants and animals. The invention of the microscope gave a great impetus to probe into the world of microcosm and get a visual knowledge about the link between one generation and the toher bringing about hereditary transmission of characters.

Harvey (1578-1657) speculated that all animals arise from eggs, with the semen only playing a vitalizing role. R.deGraff (1641-1673) opined that since the offspring has characters of both the parents, both of them should contribute to the heredity. Leeuwenhoek (1677) who studied the sperms in several animals suggested their association with the eggs.

Reproductive organs in plants were reported for the first time by Grew in 1682. It was Camerarius however who for the first time described sexual reproduction in plants. In 1717 Fairchild produced a hybrid having characters of both the parents. The hybrid was called *Fairchild's mule.*

Performation and epigenesis

The discovery of sexuality, revealed the physical link between one generation and another, thus providing a physical basis for the transmission of heredity from parents to progeny.

The homunculus theory of Wammerdam suggested that the development of an organism is a simple enlargement of a minute but completely preformed individual. According to Swammerdam, the preformed individual or homunculus could be present either in the sperm or in the ovum. During 18th and early part of 19th centuries there were two groups among "Preformationists". Those who attached importance to sperm were called *animalculists*. Subsequent studies however clearly showed, that not only ovists and animalculists were wrong, but the entire preformation theory was nothing but a figament of imagination.

Epigenesis theory

Wolff (1738-1794) who opposed the preformation concept opined that, neither the egg nor the sperm had a preformed organism, but both of them have undifferentiated mass of living substance which developed into the organism after fertilization. According to this theory called epigenesis, new organs and tissues are formed *de novo* (from the mass of living substances in thegametes) due to vital forces.

Pangenesis theory

Hippocrates (400 BC) believed that reproductive material is produced from all parts of the body so that all the characters are represented in theprogeny. Aristole (350 BC) also was a believer of direct inheritence of characters.

Charles Darwin (1809-1882) was also a votary for this type of inheritance. He modified the views of Hippocrates slightly and called it the theory of pangenesis. According to this, all parts of the body productive invisible gemmules or pangenes that are carried into sex organs *via* the blood stream. There the pangenes are assembled into gametes. During fertilization, pangenes of mother and father combine to produce the characters of the offspring.

Lamarckian theory (J.B. Lamarck 1744-1829) of inheritance, often called inheritance of acquired characters opines that any new character acquired during the life time of an individual is inherited. It will be seen from the above the Charles Darwin's pangene theory is nothing but a version of Lamarckian theory.

Germplasm theory of Weismann

August Weismann (1834-1914) disproved the pangene theory with his experiments on rats. He cut the tails of rats and observed the progeny for 22 generations. In every generation the rat progeny continued to have tails in spite of the fact that the parents were tailless. If pangene theory were to be corect, the tailless parents could not have sent the tail pangene to the gametes and consequently the progeny should have been tailless.

In order to account for inheritance, Weismann proposed his germplasm theory, which is essentially correct even today. According to this theory, the body of an organism can be divided into two - **somatoplasm** and **germplasm.** Somatoplasm constitutes thevegetative body of the organism, while the germplasm constitutes the reproductive part. Somatoplasm develops newly in every individual and there is no continuity; while the germplasm is continuous and links the generations. With the information present in germplasm, every individual develops its own somatoplasm. Hence any change or variation occuring in somatoplasm cannot be transmitted to the next generation, while those of germplams can be inherited. The fermplasm theory provided a convincing physical basis for distinction of variations into heriditary and environmental (see later in the chapter).

Variations

Variations appearing in organisms are of two categories. *Hereditary variations* and *environmental variations.* Before the classical explanation of Wismann (germplasm and somatoplasm) it was baffling the biologists as to how variations of some kind are inherited while others are not. The explanation of Weissman that the body consists of germplasm and somatoplasm and only the former constitutes the link between one generation and the other provided for the first time a basis to classify the variations into two categories (named above).

Hereditary variations

These arise in the germplasm and consequently can be passed on to the next generation. These are also called heritable or permanent variations as they go from one generation to the other. It is primarily the hereditary variations that are responsible for the variations between one species and theother.

Environmental variations

These arise in thesomatoplasm due to the influence of environmental factors and consequently cannot be passed on to the next generation. They are called transient variations becausethey will last only for one generation.

Between the two types of variations, from the evolutionary point of view, heritable variations are of some significance as they tend to carry the variation for several generations.

Some interesting experiments have been made by Clausen *et al* in the Californian (USA) plant *Potentilla glandulosa,* a member of Rosaceae, concerning the difference between heritable and environmental variations. It is generally known that the individuals of a plant species growing at diferent altitudes ina mounainous region tend to show morphological difference. The individuals that grow at sea level have long stem and leaves and take nearly six months to produce flowers from the seedling stage. A much taller form is encountered at about 4000-45000 ft above sea level and it takes only 50-60 days to flower in its natural habitat. At about 10,000 ft above sea level (alpine region) there is a dwarf form of *Potentilla glandulosa* which takes about 50-55 days to produce flowers. To what extent these differences encountered are due to environment and to what extend due to heredity? In order to find out an answer, the plant from each locality was cut into there pieces and replanted in three experimental gardens at coasta level, at 4600 feet and at 10,000 feet. The alpine race (form) grown at sea level resembles the original race, but grows to a taller height at 4500 ft. The ming altitude race (4500 ft) becomes dwarf at the coastal level and extremely dwarf at the alpine region. The coastal race grown at 4500 ft altitude is dwarfer and at 10,000 ft the plant dies. The above experiment clearly illustrates that the members of the samestock (clone) react differently to different environments as shown in the case of height. Hence variations of this type are called environmental variations.

Genotype - Phenotype

An individual organism has tow sets of attributes-an endowment which it gets from the parents and an individuality which reflects a reaction of the hereditary potentiality with the surroundings. In order to clearly demarcate the two in genetic terms, Hohansen (1909) a Danish scientist proposed the *genotype - phenotype concept.* Genotype of an individual has been defined as thesum total of all the hereditary potentialities that the organism inherts from the parents; while the phenotype is the morphological and physiological attributes of the organism - like the size shape, colour, metabolic reaction rate etc. Phenotype reflects theinteraction between the genotype and the environment. Since the individuals of a species may grow in different environments their interactions are different. In other words the same genotype may interact variouslyin different environment and produce various phenotypes. In the case of *Potentilla glandulosa,* the same genotype behaved differently in different environments.

To sum up, we can say that as far as an individual is concerned its genotype is permanent (barring extreme circumstances) while its phenotype is transient. Individuals with different genotypes some time show similar phenotypes in a similar environment.

The diverse phenotypes possible for a given genotype under various environmental influences (without causing lethality to the individual) is called the *norm* or *reaction range.* The norm or reaction range extends from one extreme to the other (with reference to all factors or parameters of the environment) to theother (with reference to all factors orparameters of theenvironment) in which a given genotype can interact and survive. The full reaction range of a genotype perhaps can never be understood in its entireity, because in order to assess this, one has to know all parameters of the environment. This is easier said than done.

Hereditary diseases

These are inborn diseases that the individual has in its genetic mechanism. It reflects an inbuilt fault in the genetic machinery. Many of the human diseases are known to be hereditary - Taste blindness to PTC, colour blindness, Diabetes, albinism, Haemophilia, Xeroderma, phenyl Ketonuria, Alkaptanuria, Sickle cell anaemia are all due tto a fault in the genetic machinery and an individual suffers from these whether one likes or not. But there is a distinction in these diseases. Some of them like Haemophilia, colour blindness etc., will affect the individual irrespective of age, physiological condition etc of the individual. On the other hand, diseases like diabetes (not all types of diabetes), even though hereditary may not affect the individual until quite late in life. This means that hereditary diseases also interact with the environment and express themselves only in a suitable environment.

In two other human diseases

Albinism and xeroderma, the former is recognised by complete loss of pigments in skin, hair, iris of the eyes etc; the latter is also a skin ailment in which exposure to sun rays results in freckles on the skin. But both these can be withstood. Albinos can protect themselves by not exposing too much tgo light, the same is the case with xeroderma people. Till now there has been no total cure for a hereditary disease - the remedies are only pelliative. But recent techniques of genetic engineering have brought a ray of hope that the faulty genes can be identified and corrected very early. However this is still in the experimental stage.

Phenocopy

In order to cure a hereditary disease or rectify a genetic abnormality; it will be necessary for us to know the norm or range of reaction of the genotype in all the environmental conditions. Even though it is difficult to fully analyse the entire reaction range of a genotype it has been possible to do so to some extent atleast in some organisms.

In the fruit fly *Drosophila melanogaster,* the normal body colour is brown with black margins. A hereditary mutant for this fly was discovered by Morghan (1`910) where the body colour was yellow. This was a genotypic character in that the body colour was constant in both the flies in all environments. In 1939, Rapport discovered that, if larvae of these flies are fed on silver salts, they develop into yellow flies irrespective of the fact whether they were genetically brown or yellow. The yellow bodies fly which is genotypically brown is a variant of the original yellow bodies fly. This is now called a phenocopy. It is easy to distinguish between the original yellow body and its phenocopy. The offspring of a phenocopy, would have brown body, if they are not fed again silver salts. The temporary development fo body colour (yellow) in brown flies is an example of norm of its genotype.

MICROBIAL GENETICS

Gentetics studies during the early post mendelian period were concentrated mainly on higher organisms. The reasons for this are not far to seek. Manipulation of organisms for breeding and study of the chromosomes can be easily carried out in higher organisms. Further, phenotypic variations are discreet and offsprings after mating could be easily analyzed for genetic varaitions. Visual observation of segregation of characters also can be done easily. In microorganisms however the picture is totally different. Organisms themselves are too small to handle. Further, when the entire organism is made up of a single cell what sort of a phonotypic variation one can visually observe in microbes? Genetic studies in microbes indeed call for an entirely different approach than the conventional methods used for higher organisms.

In microbes where no morphologically variable traits are available within the individuals of a species for the purpose of hybridization and genetic analysis, the physiological traits are the only dependable characters. What is a physiological trait? When microbes such as bacteria are cultured on a medium continuously, they undergo mutation and this can be detected by their nutritional requirements. For instance if a bacterial colony does not require the addition of biotene to the medium even though it is required for its growth, the assumption is, it is able to synthesize biotene on its own. If such a colony sometimes does not grow on a medium which does not have specially added biotene the conclusion is, the colony has lost the ability to synthesize biotene due to mutation. Such a colony can we grown on a medium which has biotent in it. By this analysis we now have two genetic variations of a bacterial colony (I) colony that can synthesize biotene and (ii) colony that can not synthesize biotene. These two can now be used for genetic studies as they have two 'phenotypic traits'. The traits however are not morphological but nutritional. In other words the two colonies mentioned above are biochemical variants. In microbial genetics the phenotypic traits are nothing but biochemical traits and genetics of microbes is essentially biochemical genetics.

Beginnings of biochemical genetics

Earliest cases of biochemical mutants in metabolic pathways were studied by A.E. Garrod in 1909 which has been published in his classical book *'Inborn errors of metabolism'*. He studied three abnormalities associated with the metabolic breakdown of phenylalanine. The three abnormalities are ***Phenyl ketonuria, Alcaptonuria*** and ***Albinism.***

The next significant step in biochemical genetics pertaining to microbes was that of G.W. Beadle and E.L. Tatum who were awarded Nobel prize for their outstanding work on the biochemical genetics of fungus *Neurospora*. Beadle and Tatum discovered several biochemical mutants in *Neurospora* and proposed their celebrated concept - **one gene, one enzyme hypothesis.** According to this hypothesis, if a biosynthetic pathway involves several steps, each step is controlled by a specific enzyme, which in turn in governed by a particular gene. Mutation in any one of the genes would alter the biochemical pathway. In addition, elucidation of biosynthetic pathways also will be made easy by the study of mutations. This is particularly true for microbes. For the study of biochemical mutations and mutations that bring about phage resistance or drug resistance, microbial systems are admirably suited. Large populations of microbes can be handled easily and mutations may be detected. Thus biochemical genetics of microbes came to be recognized as microbial genetics.

Genetic notations used for microbes. In order to discuss genetics one must be able to distinguish phenotypically one cell from the other. This is hardly possible in microbes as they do not manifest any morphological traits.

In microbes a phenotype constitutes a biochemical trait. Thus a cell which can synthesize methionine is called Met^+ and if it cannot synthesize methionine it is called Met^-. Obviously a Met^- cell must have a defective (mutated) gene which prevents the synthesis of methionine.

A typical microbe such as a bacterium has only one set of genes; it is haploid. Enkaryotic organisms on the other have two sets of genes i.e., they are diploid. A diploid cell might have normal copy of the gene and one mutated copy. In this case the heterozygote will have both copies and the gene symbols are separated by a diagonal line Met + Met^-. Alternatively a cell may have both defective copies or both normal copies. To summarize, the following conventions are used (Friefelder, 1987).

1. Abbreviations of phenotypes contain three Roman letters (the first capitalized) with a superscript + or - to denote the presence or absence of a designated character and a non superscript + or - for drug resistance.

2. The genotypic designation is always lower case, except for superscripts, all components of the symbol are itacilized.

The above conventions, however has not been adopted for bacteriophages, for which one and two letter symbols are widely used.

Some milestones in the History of Genetics

Name	Year	Contribution
Empedocles-		Semen theory
Aristotle	200BC	Semin is highly purified blood
Harvey	1651	Living beings originate from egg.
Hartsoikes and	1694	Supported preformation theory
Datempotius	1692	"Observed" miniature human being in a sperm's head
Charles Darwin	1809-1882	Pangenesis theory
August Weismann	1889	Germplasm theory
Gregor Mendel	1822-1884	Laws of Heredity; regarded as father of Genetics
Hertwig	1875	Discovered fertilization
Flemming	1879	Described longitudinal splitting of chromosomes
Benden E.W.	1883	Gametes of round worm have half the number of chromosomes as the body cells
Waldeyer	1888	Coined the term chrosome
Tschermark Devries and Correns	1900	Rediscovering of Mendelian laws
Landsteiner	1900	Blood groups
Devries	1901	Mutation in Oenothera
McClung	1902	Sex chromosomes
Bateson. W	1906	Coined the term genetics
Bateson and Punnet	1906	Linkage in peas
Johanssen	1909	Coined the word gene
T.H.Morghan	1910	Sex linkage in Drosophila
Bridges and Tatum	1941	Biochemical genetics in Neurospora
Stern	1931	Cytological proff of crossing over
Avery, Mcclead and McCarty	1944	DNA - genetic material
Muller	1946	Radiation genetics
Chargaff	1950	Base pairs are equal in proportion in DNA
Watson and Crick	1953	Molecular model of DNA
Otchoa and Kornberg	1959	DNA synthesis
Jacob and Monad	1961	Operon concept
Nirenberg, Holley		

and Khorana	1968	Genetic code
Calude and Palade	1974	Ultra structure of cells.

2

CHEMICAL BASIS OF HEREDITY

The cells have diverse chemical components of which most were known to scientists by the first quarter of the nineteeth century. These are classified into three major groups namely–carbohydrates, proteins and lipids. Of these three–proteins are the most complex and diverse and many biologists thought that they may constitute the basis of biological diversity.

Eventhough another group of macromolecular compounds – nuclei acids were discovered in 1869 by Sweedish biochemist Fredrich Miescher, it was only in 1952, their (nuclei acids) role in inheritance was proved.

Meischer collected pus filled hospital bandages and from them he isolated and purified nuclei. On chemical fractionation he found that a cellular chemical content had a very high component of phosphorous compound than what was known before. As the new compound was obtained from nucleus, Miescher named it *nuclein.* Subsequently, nuclein was isolated from many sources and due to its acidic properties, it was named nucleic acid. Upto 1952, however, interest in nuclei acids was limited to only its physiochemical studies. When its role in inheritance was proved nucleic acid became the focus of intensive research. Comparative studies of nuclei acids and proteins revealed that while proteins are of diverse types, there are only two types of nuclei acids. It was hard to believe that nuclei acids which apparently had no diversity could be the basis of heredity and not the proteins.

Presently, however, there is not even an iota of doubt as to the fact that nucleic acids alone constitute the hereditary factor. Three major experiments helped provide an evidence for this.

Bacterial Transformation

Fred Griffith, an English bacteriologist, worked with *Diplococcus pneumococci* and found out that there are two strains in this – *Virulent and Avirulent.* When injected to experimental mice the virulent strains cause the death of the mice, while avirulent strains do not casue any harm. Griffith showed that heat killed virulent strains injected into the mice along with avirulent strains cause the death of the mice and from the bodyof the mice live virulent strains can be obtained. Dawson and Sia (1931), showed that the heat killed virulent strains can transform avirulent strains on a nutrient broth. Alloway (1933), showed that even a lysed cell of virulent strain retains the ability of transformation thereby indicating that some component of the cell rather than the entire cell was involved in transformation.

Nature of the transforming agent

Three scientists – O.T. Avery, Colin Macleod and Maclyn McCarty (1944), went into serious search of the transforming agent from the lysed cells of bacteria. Even after digestion of polysaccharide coat, the cell retained the ability of transformation. This shows that the capsule has no role in transformation. Similarly, protein degradation also had no adverse effect on transformation. With the two important components out of the way, it thus became clear that nucleic acids perhaps might play a role in transformation. It was shown that the DNA fraction of nucleic acids was responsible for transformation. It was also shown that if the cells are treated with DNAse (a DNA degrading enzyme), the transforming ability will be lost. Enzymes which digest RNA do not disturb transformation.

Bacteriophages and chemical basis of heredity

Bacteriophages are viruses that attack a bacterial cell and multiplty inside the (bacterial) cell. The phage ataches itself to the bacterial cell and insects its components into the cell, where replica of phages are produced. Hershey and Chase (1952) conducted an experiment to find out which component of the bacteriophage carries the genetic information necessary for the replication of viruses. They cultured the host bacterium *Escherichia coli* in a medium containing radio isotopes of Sulphur (S^{35}) and phosphorus (P^{32}). When T_2 phages attacked these cells subsequently, the labelled compounds get incorporated in phage DNA which always has radioactive phosphorus but not any sulphur. In the next set of experiment only the protein coat of T_2 phage was labelled with S^{35} and the DNA core of another was labelled with P^{32}. Hershey and Chase separated the radioactive phages by centrifugation and allowed them to infect a population of non radioactive bacteria. A few minutes after initial infection, they separated the phages from bacteria by agitating the medium. After centrifugation, the host and the parasite were separated by centrifugation and their radioactive measured. About 95% of P^{32} was found in bacteria while about 95% S^{35} was seen in phages. This experiment conclusively proved that during infection, only DNA is injected into the medium. This is the reason why the non radioactive bacteria showed labelled P^{32} (as they obtained it from labelled phage DNA). For this work, Hershey was awarded Nobel prize in 1969.

Thus, the experiments of Hersehy and Chase, Avery, Mcleod and McCarty. Griffith etc., clearly proved that the DNA constitutes the chemical basis of heredity.

TMV and Chemical Basis of Heredity

The discovery that DNA is the hereditary material posed some problems to explain the inheritance in instances such as TMV (Tobacco mosaic virus), where there is no DNA (TMV is a RNA virus). But, the absence of DNA has not prevented TMV from transmitting its characters from one generation to the other. TMV is made up of a RNA core and a helical protein coat. The protein RNA component of TMV can be easily separated by shaking the viral particles in a mixture of water and phenol. Frankek Conrad and Wichams (1955), separated the protein and RNA from TMV and reassembled the particles from the individual components thus indicating the purification process (of the components) does not affect their function. Subsequently, Gierr and Shramm (1956), showed that on contact with the host only the RNA can bring about the disease and not the protein coat. In TMV therefore, it is RNA as the hereditary material. It can therefore be concluded that nucleic acids in general, and DNA in particular is the hereditary material in most of the organisms; and in the absence of DNA (as in some viruses) RNA can constitute the genetic material.

Nucleic Acids

These are the most important constituents of the nucleus playing a vital role in the inheritence of characters from one generation to the other. There are two types of nucleic acids – Deoxyribose nucleic acid (DNA), and Ribose nucleic acid (RNA). While both constitute the genetic material in different organisms, it is predominantly the DNA which is the genetic material in most of the organisms. The following are some of the important milestones in the discovery of Nucleic acids.

1. Meischer (1869), isolated nucleic acids and called them nuclein.
2. Hertwig (1884), suggested a heriditary role to the nuclein
3. Kossel (1884), recognised purines and pyrimidines in the nucleic acids.
4. Feulgen (1912), devised basic fushsin dye positive only to DNA.
5. Astbury and Bell (1938) showed that the bases in DNA molecule are arranged one above the other.
6. Avery, Macleod and Macarty (1944), presented evidence for DNA to be the genetic material.
7. Chargaff (1950), showed that DNA contains equal amount of purines and

pyrimidines.

8. Wilkins *et al* (1950), identified the molecular distance between the bases.
9. Furberg (1952); DNA is a coil of single chain.
10. Pauling and Corey (1952); DNA is a helix of three chains.
11. Watson and Crick (1953) demonstrated the double helix model for DNA molecule.
12. Meselson and Stahl (1958), demonstrated the semi conservative replication in DNA.
13. Jacob and Monod (1961), postulated the presence of RNA in protein synthesis.

DEOXY RIBOSE NUCLEIC ACID (DNA)

DNA is the main genetic material in almost all organisms except for certain viruses. In prokaryotes like *Escherichia coli,* the genetic is a single giant molecule of *DNA*. It is mainly concentrated in the nucleus, it has also been reported in cytoplasmic organelles like chloroplast, mitochondria etc.

Chemical Composition of DNA

DNA extract from many living organisms have revealed it to be a complex macromolecule. It is a long chain polymer cmposed of monomers (units), called nucleotides. Each nucleotide inturn is composed of three sub units. These are – (a) nitrogen base, (b) sugar and (c) phosphate.

Nitrogen bases

The nitrogen bases are of two kinds viz., pyrimidines and purines. Both pyrimidines and purines are heterocyclic compounds, with a basic six membered ring in which one or the other carbon position is replaced by nitrogen.

The pyrimidines have a basic six membered ring of benzene in which, position 1' and 3' are replaced by nitrogen. There are two kinds of pyrimidines – *Cytosine* and *thymine.* Structurally, cytosine has a single six membered ring with a hydroxyl group at the 2' position and an amino group at the 6' position *(2 hydroxy, 6 amino pyrimidine).* Thymine also has a single six membered ring with hydroxyl groups at the 2' and 6' position, and a methyl group attached to the 5' position *(2, 6 hydroxy 5 methyl pyrimidine).*

The purines have a double ringed structure. Attached to the six membered benzene ring is a 5 membered imidazole ring. The imidazole ring is joined to the hexagonal ring at 4' and 5' positions. The benzene ring has nitrogen in the 1' and 3' position (replacing Carbon).

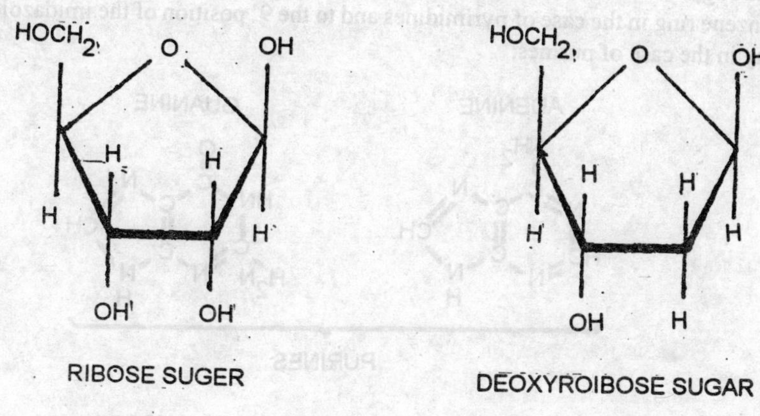

Fig. 2.1 Chemical basis of heredity
Pentose sugars of nucleic acids (left, robose; right deoxyribose)

There are two types of purines viz. *Adenine* and *guanine.* Adenine has an amino group at the 6' position (*6 aminopurine),* while guanine has an aminogroup at the 2' position and a hydroxyl group at the 6' position (*2 amino, 6 hydroxy purine).*

Sugar

The sugar molecule in DNA is basically a ribose sugar. The ribose sugar is a *Pentose Sugar.* It is a 5 membered ring. The ring consists of four carbon atoms while the 5' C atom is outside the ring. Positions 1', 3' and 5' in the ring have OH groups, while the 2' carbon position has only H. There is no O; hence the sugar is said to be Deoxyribose (no oxygen in the second carbon position).

Phosphate

It has one central phosphorus with four oxygen atoms.

Nucleosides

A nitrogen base coupled with sugars forms a nucleoside. DNA has four nucleosides –

(a) Deoxycytidine
(b) Deoxythymidine
(c) Deoxyadenosine and
(d) Deoxyguanosine

The sugar molecule is attached to the nitrogen base to the 3' position in the benzene ring in the case of pyrimidines and to the 9' position of the imidazole ring in the case of purines.

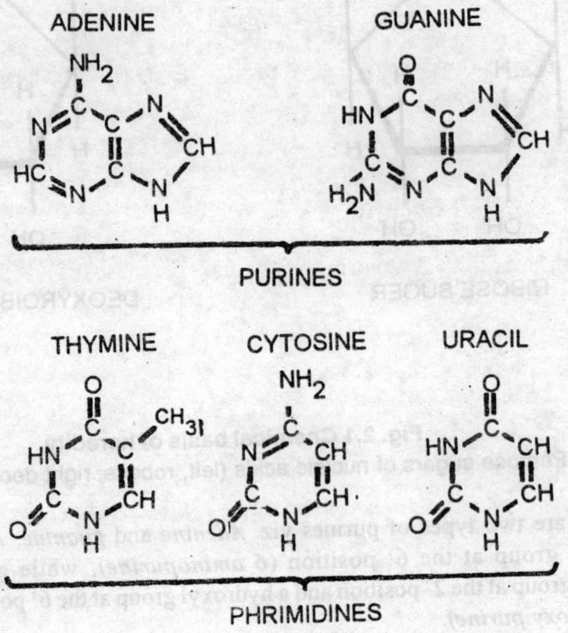

ADENINE GUANINE

PURINES

THYMINE CYTOSINE URACIL

PHRIMIDINES

Fig. 2.2 Chemical basis of heredity
Nitrogen bases of nucleic acids

Nucleotides
A nucleotide is derived from the nucleoside by the addition of phosphoric acid molecule to the latter. The phosphate group is usually linked to the two sugar molecules above and below. The linkage is with the 3'C atom of one deoxyribose sugar and with the 5' C atom of another. DNA has four nucleotides – *Deoxycytidilic acid, Deoxy thymidilic acid, deoxy adenylic acid* and *Deoxyguanilic acid.*

Polynucleotide
DNA is said to be a polynucleotide chain with a number of nucleotides joining together with the help of DNA polymerase.

Molecular model of DNA structure
A molecular model for DNA structure first proposed by J.D. Watson and F.H.C

Crick (1953) has been universally accepted. They were awarded Nobel prize for their discovery. The work of Crick and Watson has been supported by X-ray diffraction picture obtained by Wilkins and his associates.

According to this model, a DNA molecule is composed of two chains that are spirally coiled and form a double helix, much like a spiral stairway. The distance between the two chains is uniform and it is maintained by base pairing. The back bone of the chain is made up of alternating sugar and phosphate molecules. At specific intervals, the chains are joined by 'steps'. Each step is made up of two nitrogen bases. Each nitrogen base is attached to the sugar molecule of the opposite chain. Between themselves, the nitrogen bases of a set are joined by hydrogen bonding. The pairing of the nitrogen bases is always specific. It is made up of a purine and a pyrimidine i.e., a double ring and a single ring. Two purines would occupy too much space and can not be accomodated between two chains; similarly two pyrimidines would form a short step and leave a gap between two chains. Hence the pairing is always between a purine and pyrimidine. Because of this pairing,in a DNA molecule, the total purine content is always equivalent to total pyrimidine content.

The purine, pyrimidine base pairing is further specific. It is always Adenine with Thymine and Guanine with Cytosine. Adenine and Thymine are joined by two hydrogen bonds through atoms attached to position 6' and 1'. Cytosine and Guanine are joined by three hydrogen bonds at positions 6', 1' and 2', the bonding to hydrogen is weak; but this facilitates easy separation of two strands during replication. According to Crick and Watson, the helix has a diametre of 20 A^0 and there is one complete turn to the helix at every 34 A^0 and has a stack of 10 nucleotides at every turn.

One unique character of DNA is its unity and yet infinitismal variety. The DNA structure mentioned above is uniform in all organisms, whether it is man or virus. At the same time it is also true that DNA is the genetic material that decides the characters of individual organisms. Then how does DNA decide the individuality? The answer lies in the base pair sequence. The combination of A/T, G/C is uniform, but their sequence is not; for instance in one individual it may be AT, G/C, GC, AT, GC< AT etc. There is no limit to the variety in which base pair sequences are arranged.

This is the secret of individual differences. In no two individuals, the base pair sequence is same.

Unusual bases in DNA

The normal bases in DNA are A, T, G and C. In some viruses like PBS1 and PBS2, uracil replaces thymine. In some bacteriophages, C is replaced by 5 hydroxy methyl cytosine.

Fig. 2.3 Chemical basis of heredity
F.H.C. Crick (left and J.D. Watson (right)

Single strand DNA

Although DNA in most organisms is double stranded, in some phage viruses (ϕ x 174), single stranded DNA is found. The following table summaries the differences between double stranded DNA and single stranded DNA.

	Character	Double Strand DNA	Single Strand DNA
1.	U.V. Absorption	Constant from 10-80^{0}C and then rises rapidly	Steady increase from 20^0 to 90^{0}C
2.	Action of formaldehyde	Resistant	Not Resistant

3.	Ratio of	equal, A=T and 1.33:0.98:0.75	A:T:G:C is
4.	Shape	Linear	Circular
5.	Respiration	Two strands separate and replication begins	Becomes double (replicativeform) temporarily.

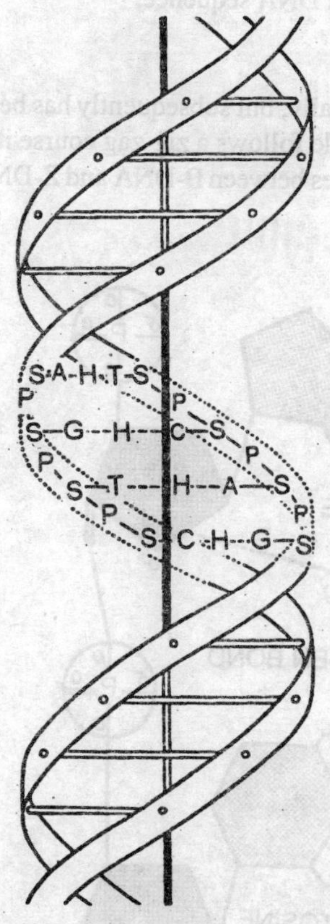

Fig. 2.4 Chemical basis of heredity
The double helix model od DNA

Alternative forms of DNA double helices

The structure of DNA molecule described above as visualized by crick and Watson assumes that the helical coiling of the molecule is right handed and this form of DNA has been called the B-form. Subsequent researches, however, have indicated that other types of coiling are also possible for the DNA helex. These alternative forms of DNA differ from one another in the following respects.

1. Number of residues per turn (<u>n</u>), and
2. Spacing of residues along the helix (<u>h</u>)

The existence of DNA as double helix has also been confirmed by experiments used for measuring the number of base pairs per turn of the helix and this works out to be 10.4 instead of 10 in the B-DNA. This leads to the modification of angle of rotation from 36 to 34.6. Thus, it is possible that there could be structural variants of DNA with varying values for <u>n</u> and <u>h</u>. Three structural variants of DNA A,B and C are known since a long time and transitions between them are also possible. The following table gives a brief summary of the characters of these forms of DNA.

Helix type bryophytes	Condition	Basepair	Rotation per
A	75%	11	+32.7 (right handed)
B	92%	10	+36.0 (right handed)
C	66%	9.33	+38.6 (right handed)
Z	very high salt	12	-30.0 (left handed)

The forms of DNA are assumed to be found in all molecules depending on the surrounding conditions. In addition to the above, D and E forms of DNA are also found very rarely with only 8 and 7.5 base pair per turn. These are found only in some molecules of DNA lacking in Guanine. In contrast A,B and C forms are found in all molecules irrespective of DNA sequence.

Z-DNA

This type of DNA was first synthesized artificially, but subsequently has been reported from living cells also. As this molecule follows a zig-zag course it is called Z-DNA. The resemblances and differences between B-DNA and Z-DNA are as follows –

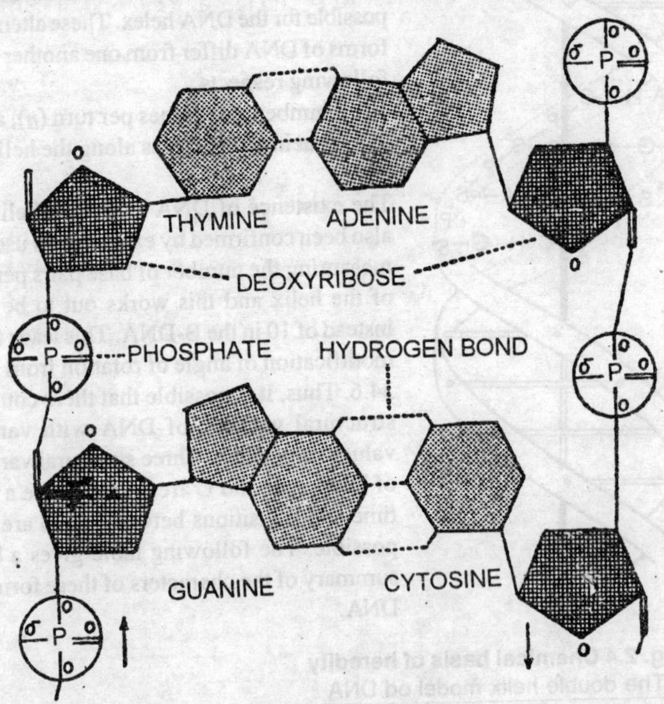

Fig. 2.5 Chemical basis of heredity
Arrangement of nucleotides in a segment of the DNA model

Resemblances between B-form and Z-form

1. Both are double helical.
2. The two strands are antiparallel in both forms.
3. Both forms exhibit A=T and G=C pairing.

Differences between Z-DNA and B-DNA

1. Z-DNA is left handed, while B-DNA is right handed.
2. The phosphate back bone is regular in B-DNA, while it is zig in Z-DNA.
3. In Z-DNA, sugar residues have alternating orientations as a result of which repeating, units are dinucleotide, whereas in B-DNA they are monocleotides.
4. Z-DNA has 12 bp for every twist of 360^0 while B-DNA has only 10bp.
5. As more base pairs are accommodated in one helix in Z-DNA, the angle of twist is 60^0 as against 36^0 in B-DNA.
6. One complete helix is $45A^0$ in Z-DNA, while it is $34A^0$ in B-DNA.
7. The diameter of the helix is $18A^0$ in Z-DNA, while it is $20A^0$ n B-DNA.

The role of Z-DNA is not certain, even though it is thought to play a role in regulation of gene activity, recombination or rearrangement of DNA.

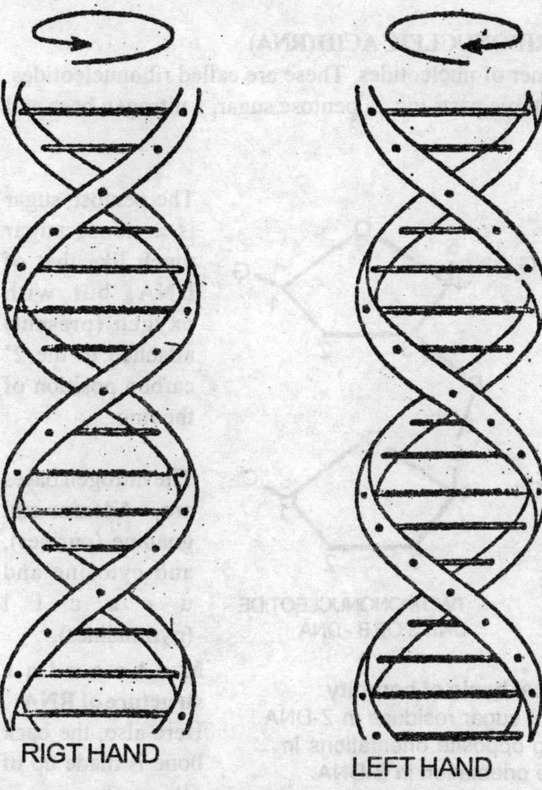

Fig. 2.6 Chemical basis of heredity
Two helices, one showing right handed coil as found in B-DNA and the other showing left handed sense as found in Z-DNA.

RL Model

A new model for the molecular structure of DNA has been proposed by Sasisekharan and associates (1976-89), working at the Indian Institute of Science, Bangalore. According to them, the right handed (R) helix of B-DNA cannot easily explain the unwinding of the coil due to topoligical differences. Hence they suggested a combination of right handed (R) and left handed (L) helices, which they called the RL model. In this model, a B-DNA molecule will have alternating left handed and right handed segments of approximately 5 base pairs in a repeat of 10 base pairs. This model provides conformational flexibility. Many molecular biologists opine that there is a strong experimental evidence for the RL model although one is not sure of the continuous length of right and left handed segments in a molecule of DNA.

Super Coiling of DNA

The double helix structure of DNA represents in linear molecule, but in living conditions the molecule is a closed structure without free ends. In bacteria, the DNA is circular while in eukaryotes also large loops appear to represent part of a circle. As a result of this, the DNA becomes supercoiled. One supercoil is introduced everytime when the duplex is twisted about its axis. The super coil may be negative (winding opposite to the helix coiling), or positive (winding in the same direction as the helix coiling).

RIBONUCLEIC ACID (RNA)

The RNA is also a polymer of nucleotides. These are called ribonucleotides. Each ribonucleotide has three parts viz., a pentose sugar, a nitrogen base and a phosphate group.

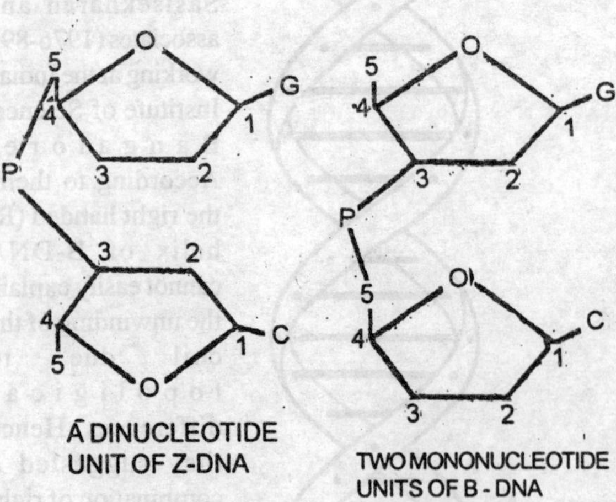

Fig. 2.7 Chemical basis of heredity
Orientation of adjacent sugar residues in Z-DNA and B-DNA showing opposite orientations in Z-DNA and same orientation in B-DNA. This results in dinucleotide units in Z-DNA as against mononucleotide units in B-DNA

The pentose sugar is a ribose sugar much like that of DNA, but with oxygen (present) attached to the 2' carbon position of the ring.

The nitrogen bases are – Adenine and guanine (purines), and cytosine and uracil (pyrimidines).

Molecular structure of RNA

Here also, the back bone is made up of alternating sugar and phosphate. The single chain folds to enable the base pairs to meet here and there to form the steps. The pairing is between A-U and C-G. Occasionally, the RNA may be double stranded but does not form a helix like DNA.

Single stranded RNA is the genetic material in plant viruses, poliomyletis viruses etc. Double stranded (non helical) RNA occurs as the genetic material in reoviruses.

Types of RNA

The following are the types of RNA found :

(a) Genetic RNA
(b) Ribosomal RNA
(c) Messenger RNA
(d) Soluble or Transfer RNA.

Genetic RNA

In certain viruses, RNA is the only genetic material. DNA being absent. This RNA is self- replicating; it is called RNA dependent RNA synthesis. The viral RNA directly functions as a messenger and produces RNA polymerase enzyme as well as proteins.

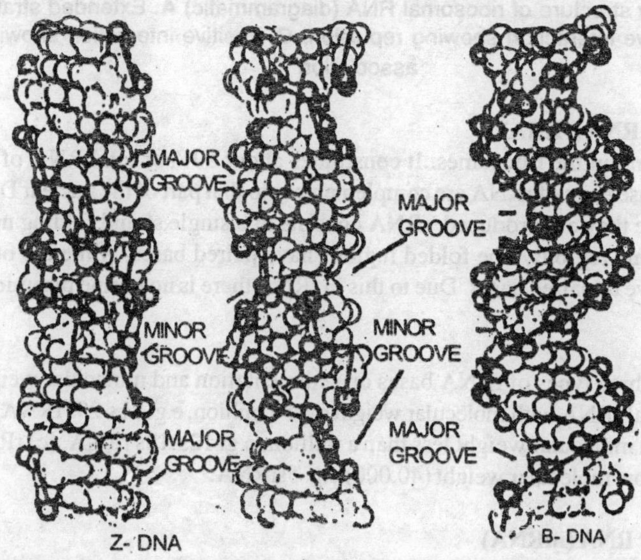

Fig. 2.8 Chemical basis of heredity
Side views of A-DNA and B-DNA. Two views of Z-DNA are 300 apart. Irregularity of backbone in Z-DNA is shown by heavy lines showing the path of phosphate residues, which is quite regular and smooth in B-DNA

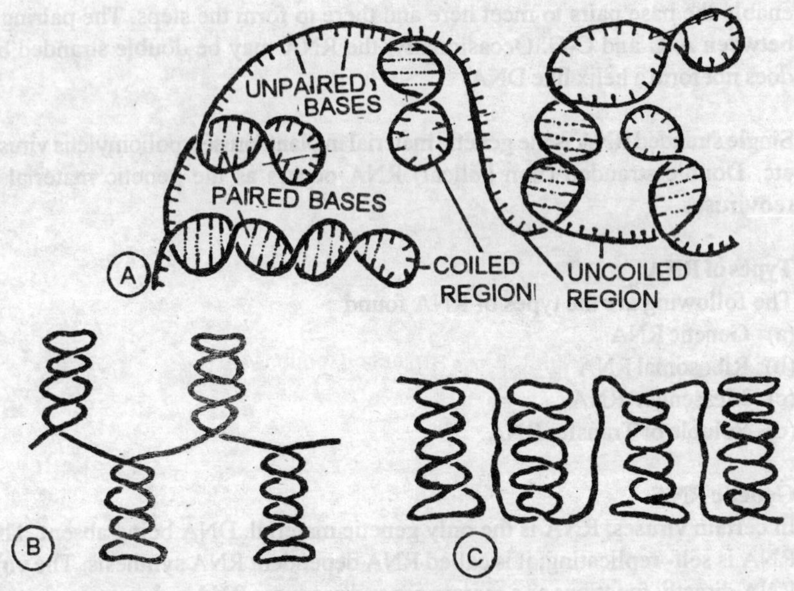

Fig. 2.9 Chemical basis of heredity
Secondary structure of ribosomal RNA (diagrammatic) **A.** Extended strand, **B.** Negative interaction showing repulsion, **C.** Positive interaction showing association

Ribosomal RNA (rRNA)

This is found in the ribosomes. It constitutes about 80% of total RNA of the cell. The base pairs of rRNA are complementary to that part of the nuclear DNA from where they are produced. rRNA consists of a single strand folding upon itself at some regions. The folded regions have paired bases, while the other regions have unpaired bases. Due to this in rRNA there is no purine pyrimidine equality.

There are three types of rRNA bases on sedimentation and molecular weight. These are (a) rRNA with molecular weight over a million, e.g. 21s-29s rRNA, (b) rRNA with molecular weight less than a million, e.g. 12s-18s rRNA, (c) rRNA with very low molecular weight (40,000) e.g. 5srRNA.

Messenger RNA (mRNA)

mRNA carries the message from DNA to ribosomes during protein synthesis. It is about 3-5% of total RNA content of the cell.

mRNA has a molecular wieght of about 5,00,000 with a sedimentation cefficient of 8s. This, however, is highly variable mRNA and is always single stranded and there is no base pairing, even though random coiling may be observed. The mRNA molecule is produced from the coding strand and its base sequence is complementary to the coding strand. The following are some of the general features of mRNA :

1. *Cap :* This is a blocked methylated structure found at the 5' end of RNA.
2. *Non coding region :* This follows the cap and consists of 10–100 nucleotides. There are no codes for proteins in this region.

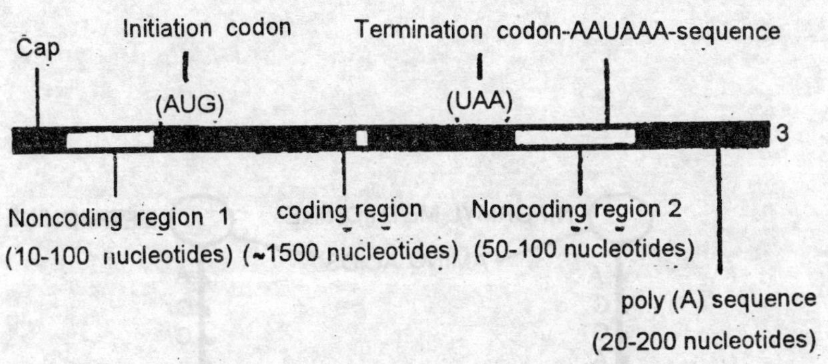

Fig. 2.10 Chemical basis of heredity
General features of eukaryotic mRNA

3. *Initiating codon :* This region begins the coding.
4. *Coding region :* Codes for proteins; and has about 1500 nucleotides.
5. *Termination codon :* The coding region is followed by a terminating codon, which stops translation.

Transfer RNA (tRNA or sRNA)

The transfer RNA also called soluble RNA, is next in quantity to rRNA. It has a low molecular weight of about 25,000–30,000 and the sedimentation co-efficient is 3.8s. It is made up of about 70–90 nucleotides.

tRNA is synthesised in the nucleus on a DNA template. Only a minute fraction of DNA (0.025%) codes for tRNA. The tRNA molecule consists of a single strand looped around itself with the 3' end always terminating in a C-C-A sequence. Some bases are paired.

The function of tRNA is to carry aminoacids to mRNA during protein synthesis. Each aminoacid couples with a specific tRNA. Since the protein molecule consists of about 20 types of amino acids, there must at least be 20 types of tRNAs. But there could be many more, sinced there are two tRNAs for every amino acid in several cases.

Secondary structure of RNA

This refers to the arrangement of the chain of the molecule to enable it to function in a proper way. Many models have been proposed for the secondary structure of tRNA. Of these, clover leaf model proposed by Holley (1966), is most widely accepted.

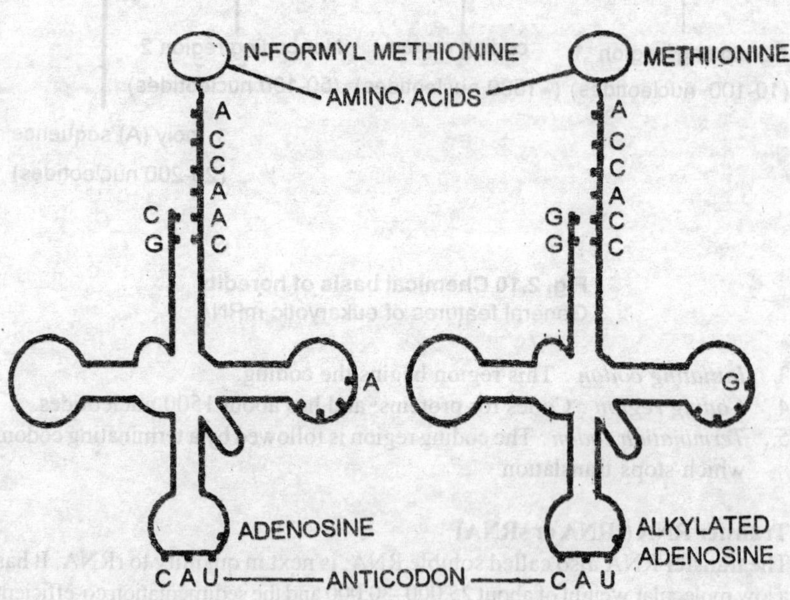

Fig. 2.11 Chemical basis of heredity
Secondary structure of tRNA (Clover leaf model)s

According to the clover leaf model, the single chain of tRNA is folded on itself to form five arms. As a result of this folding, the 3' end and 5' end lie adjacent to each other. Each arm consists of a basal stalk, and a loop. The double stranded stalks show the typical base pairing. There could be an exceptional mismatch. The loops do not have base pairing. One of their arms which does not have a loop is called the acceptor as this is the binding site for amino acids. The second arm is called the DNA arrm. This has a synthetase site which recognises the amino acid activating enzyme. The fourth is a variable arm; it has a stem, but the loop may or may not be present. The Tψ arm has a ribosome recognition site.

There is also a tertiary structure to the rRNA. This is brought about by H bonding between bases, and between bases and sugar phosphate backbone.

3

GENES AND CHROMOSOMES

The coloured thread like structure that appear during cell division have hidden within them all the secrets of genetic characters that are passed on from one generation to another generation. Housing the particles of inheritance, the genes, these vehicles of heredity are so important for all organisms; because if it these that delimit one individual from another. These are the chromosomes (Gr. *Chromo* = coloured, *soma* = body) which appear like coloured threads on staining. In reality however these are not coloured. In this chapter we will learn about the structure and functions of these unique threads of life.

History

It was Waldeyer (1888) who first coined the term chromosome to the darkly staining bodies of the nucleus. Fleming (1879) actually had described the splitting of the chromosomes, but he had coined ther term chromatin to refer to them. Sutton and Boveri (1902) usggested that the chromosomes are the carriers of hereditary particles. Morgan (1933) discovered the role of chromosomes in the transmission of hereditary characters.

Chromosomes occur as a network of fibrils called chromonema in the interphase nucleus. At the time of cell division the chromonema breaks and condensation takes place forming specific number of chromosomes.

Number of chromosomes

The number of chromosomes is constant for a given species. This number has been evolved during the course of evolution and characterises the species. Chromosome number and morphology is one of the parameters in deciding the phylogeny and interrelationship of the group as well as the individual. The number of chromosome in various species is as varied as the organism themselves. They range fro 4 to 500.

Polidy of chromosomes

The total number of chromosomes present in the cell of an organism is called *ploidy*. In all sexuallyreproducing higher organisms the ploidy is of two types - *haploid* and *diploid*. The somatic cells contain diploid or two sets of chromosomes (2x). In these organism during reproduction, the chromosome number would be halved in meiosis. As a result, the gametes have haploid or one set (x) of chromosomes. When two gametes fuse, the resultant zygote has two sets of chromosomes (diploid). It is this zygote which develops into the adult individual.

In lower organisms like many algae and fungi, the somatic cells have haploid chromosome numbers. In these organisms only the zygote will have diploid chromosomes. The meiotic division which halves the number of chromosomes takes place in zygote rather than during gamete formation. In some unusual cases the cells have many sets (polyploid) of chromosomes, in which case it is a multiplication of the base number called x.

TABLE

Group	Name of the Organism	Chromosome number
Protozoa	*Paramecium*	30-40
Coelenterata	*Hydra vulgaris*	32
Nematoda	*Ascaris Lumbricoides*	24
Chordata	*Rana esculenta*	24
	Homo Sapiens	26
Fungi	*Mucor heimalis*	46
Pteridophyte	*Ophioglossum spp*	2
Gymnosperm	*Pinus Ponderosa*	500
Angiosperm	*Crepis Capillaris*	6
	Allium cepa	14

Size of the chromosomes

There is a great variety in chromosome size and hspae. In size they range from 0.1μ to about 30μ in length. In exceptional cases however (polytene chromosomes) the chromosomes may be as long as 2mm. All the chromosomes in a species may be of similar size or they may vary. In *Yucca arakansana*, there are two distinct size groups. In some human beings there is a gradation in size. In some instances (*Mediola*) the cells at different regions (roots and shoot tip) show different sizes of chromosomes. In general, the chromosomes of plants are larger than those of animals. Monocotyledons have larger chromosomes than dicotyledons.

Structure of the Chromosome

The following are the parts usually associated with the chromosome.

Pellicle and Matrix

Light microscopic studies have revealed that each chromosome is surrounded by a membranous is surrounded by a membranous structure called *Pellicle.* The membrane encloses a jelly like substance called *matrix.* The presence of matrix has been shown in *Luzula campestris* (Juncaceae). Electron microscopic studies however have revealed, that there is nothing like a pellicle enveloping the chromosomes.

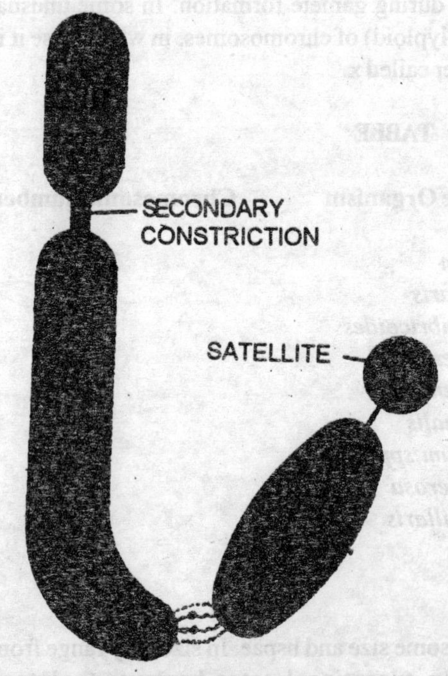

Fig. 3.1 Chromosomes
External view of a metacentric chromosome

Chromonema

When the nucleus is resting, the chromosomal material will be in the form of a network called chromonema. At the time of cell division these break up and condense to form chromosomes. The metaphase chromosomes have two thread like structures caled chromatids. The chromatids are held together at the region of primary constriction (for details see later in the same chapter). Treatment of chromosomes with a protein hydrolysing enzyme (trypsin) will reveal such chromatid to have two subchromatids; but there is no unanimity on this. Various chromosome models have proposed chromonema coils ranging from 2 to 32. The chromonema coils (subchromatids) were first observed by Baranetzky

(1880) in the pollen mother cells of *tradescantia.* It was Vejdovsky (1912) who first gave the name chromonema to these coils.The number of chromonema fibrils not only varies with different species, but at different growth stages in the same cell.

The chromonemal fibrils are coiled around each other in two different ways. These are *Paranemic coils* and *Plectonemic coils.*

The paranemic coils are loosely coiled and the threads are easily separable from one another. The plectonemic coils however are interwined very closely so that they are not easily separable.

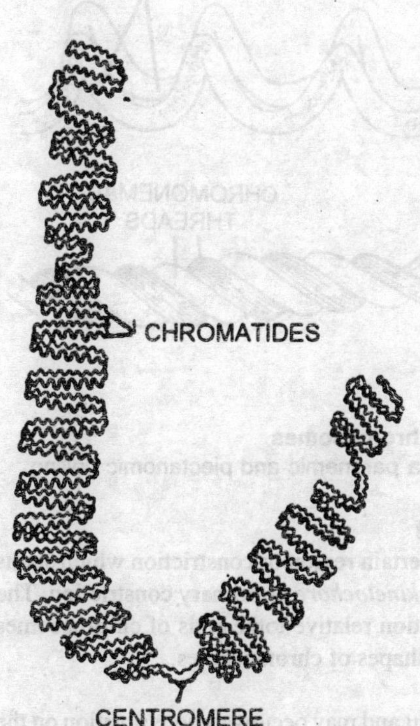

Fig. 3.2 Chromosomes
Internal structure of a metacentric chromosome

The extent of coiling in the chromonema fibrils during cell division depends on the length of the chromosomes. These coiling are of three categories **Major Coils** Each coil consists 10-30 gyres **Minor Coils** These are at right angles to the major coils and have many more gyres than major coils.

Standard or somatic coils : These are coils found in the chromonema of cells during mitosis. They are similar to major coils.

Chromomers

The chromonema in the chromosomes particularly during mitotic prophase appears to contain alternating thick and thin regions giving a bead like appearance in a chain. The thick bead like regions are called *chromomers* and the region in between the two chromomers is called the *interchromomere.* The position of chromomers in the chromonema appears to be specific to a given chromosome in an individual.

What are Chromomers

Cytologists have gien various interpretations; while some regard it as the region of condensed nucleoplasmic material, others interpret it as region of superimposed chromonemal coils. This view is supported by electron microscopic observations. Previously the chromomeres were mistaken for genes.

The chromomers were first described by Balbiani in 1876 and later by Pfitzner in 1881. Belling (1928) had equated chromomeres with the genes in liliaceous plants. But McClintock (1944) has shown that genes are located in the interchromomere region of the chromonema. However in case of polytene chromosomes the chromomeric bands are associated with certain regions.

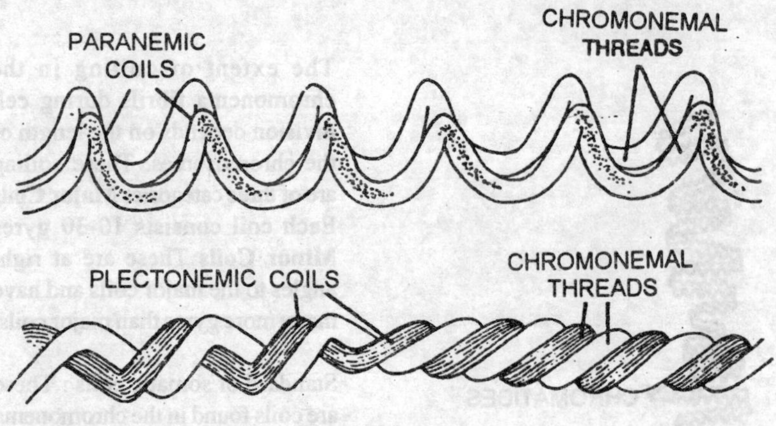

Fig. 3.3 Chromosomes

Diagrammatic representation of a paranemic and piectanomic coiling

Centromere (Primary Constriction)

The stained chromosomes reveal a certain region or constriction which stains lightly. This is called *centromere* or *kinetochore* or primary constriction. The centromeres occupy a constant position relative tothe ends of chromosomes and are responsible for the various shapes of chromosomes.

The centromere has a diameter of 0.2μ and may occupy various position on the chromosome. It (centromere) consists of spherules or granules (*kinosome*) and interchromomeral fibrillae. the numbeer of fibrillae per centromere varies. The chromonema are less coiled in the centromeric region and hence stain lightly. DNA however is present in the centromeric egion (Lime-de-Faria 1955)

Most of the organisma have their chromosomes with a single centromere (*monocentric)*. Certain other chromosomes have two centromeres (*dicentric)*.

In hemipteran insects however the chromosomes have many centromeres (*polycentric)*. Such a centromere is said to be a *diffused centromere.* The proof for centromere being diffuse comes from the fact that when such a chromosome is fragmented by X-Rays, each bit functions like a typical chromosome, replicates itself and participates in the division cycle. Diffused centromere has also been reported in *Luzulla purpurea* of the family Junacaceae (Brown 1954). According to some cytologists, chromosomes with a diffused centromere are more primitive than the ones with a localized centromere. The polycentric chromosomes found in the nematode *Ascaris megalocephala* are unique in the sense they are restricted to only those cells which constitute the germ line (Swanson 1957)

The position of the centromere in the chromosome (which is constant to a given chromosome) varies i.e., it may occupy different position. Based on this, four morphological shapes have been identified in chromosomes. These are :-

1. Metacentric

The Centromere occupies a middle position with reference to the length of the chromosome. The two arms thus resulted are almost equal in length. During anaphase movement in cell division metacentric chromosomes appear 'V' shaped. e.g. : *Trillium, Tradescantia* etc.

2. Sub Metacentric

When the centromere is located some distance away from the middle region of the chromosome, the position is said to be sub median and the chromosome sub metacentric. As a result, one arm of the chromosome will be shorter than the other. During anaphase movement, sub metacentric chromosomes appear 'L' shaped e.g : *Human beings.*

3. Acrocentric

In this case, the centromere is situated is situated exactly at one end, the chromosome will be having only one arm. Telocentric chromosomes are very rare. Truly telocentric chromosomes have been identified by Marks (1957) in certain plants, protozoa and certain mammals. Cleveland (1949) has also reported in certain protozoa inhabiting the digestive tract of wood termites, the occurrence of telocentric chromosomes. The earlier report regarding chromosome IV of *Drosophila Melanogaster,* being telocentric has been disproved as a second arm has been seen.

Acentric Chromosome

In certain instances, chromosomes may undergo fragmentation resulting in one fragment without a centromere. Such a chromosome fragment is said to be *Acentric.*

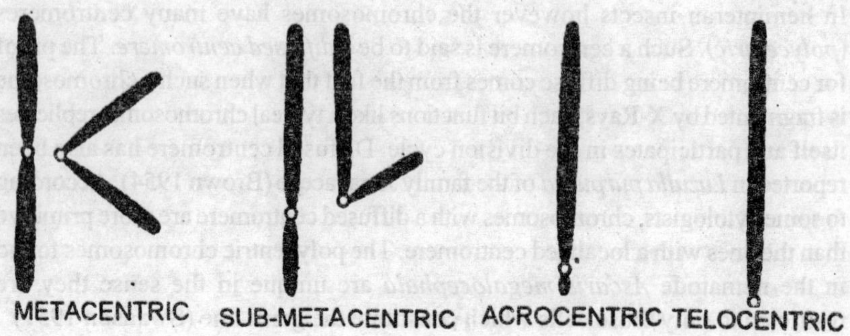

Fig. 3.4 Chromosomes
Types based on centromere position

Arm Ratio

The ratio of length between the long arm and the short arm is said to be the *arm ratio,* which is constant for a given chromosome. The arm ratio is less in metacentric chromosomes and high in acrocentric chromosomes.

Centromeric Chromomeres

The centromere should not be mistaken or equated with a chromomere as the formere stains lightly and is a constriction while the latter is a swelling on the chromosome found throughout its length. Metaphase chromosomes which have two chromatids exhibit in their centromere four granular structures. The granules called centromeric chromomeres are about 0.5μ in size and are arranged in a square.

Functions of Centromere

1. Centromere provides mobility to the Chromosome. Chromosomes lacking in centromere fail to orient themselves properly on the anaphase plate, lag behin and finally get eliminated.
2. Centromere decides the shape of the chromosome.
3. Centromeres attach themselves to the spindle fibres during anaphasic movement.
4. The centromere divides along the longitudinal axis of the chromosome at the beginning of anaphase. Very rarely the centromere may divide at right angles to the chromosome giving rise to two pieces of centromere, each bit holdng the arms parallel to each other and genetically alike (because of

duplication) as they represent the replicated parts of a single chromatid. Such chromosomes with genetically identical arms (having same genes on both arms) are called **isochromosomes** (Darlington 1939).

Secondary Constrictions

These are lightly straining areas on the chromosome other than centromere. One or more secondary constrictions may be found in the chromosomes. The part of the chromosome (arm) present beyond the secondary constriction is called a *Satellite.* There may be one or two satellites depending on the number of secondary constrictions. Chromosomes with satellites are called *SAT-Chromosomes (*sine acid thymonucleinico - without DNA). Generally in a genome atleast one chromosome will have a satellite.

Secondary constrictions associated with the nucleolus are called *nucleolar organisers.* Usually in a diploid organism two homologous chromosomes have additional constrictions where nucleolus will be located during the interphase stage of the nucleus. During prophase, the nucleolus disappears to appear again in the identical area. The nucleolar organising region is best seen during the pachytene stage and it is heteropycnotic in maize (McClintock). In polyploid organisms, nucleolar organisers will be more than two in number. Even though it ws believed earlier that the nucleolar organisers lack DNA, subsequent studies have shown that it has about 0.3% of nuclear DNA. The gent loci at the region are responsible for the formation of 28s ribosomal RNA.

Telomeres

The terminal part of chromosomes are called telomeres which have polarity, preventing the tips of other chromosomes from fusing with it. Telomeres however attach themselves to the nuclear envelope during interphase.

CHEMICAL COMPOSITION OF CHROMOSOMES

The two main chemical components of the chromosome are nucleic acids and proteins.

Proteins

Besides nucleic acid, there are three types of proteins in the chromosomes. These are *protamines, Histones* and *non histone proteins.*

Protamines are basic proteins with a molecular weight less than 4,000. They are rich in amino acid arginine and are found in the spermatozoa of salmon, herring and other fish, snails and squids. The protamines are helically bound to be DNA. Each polypeptide of protamine consists of about 29 amino acids of which 19 residues are arginine.

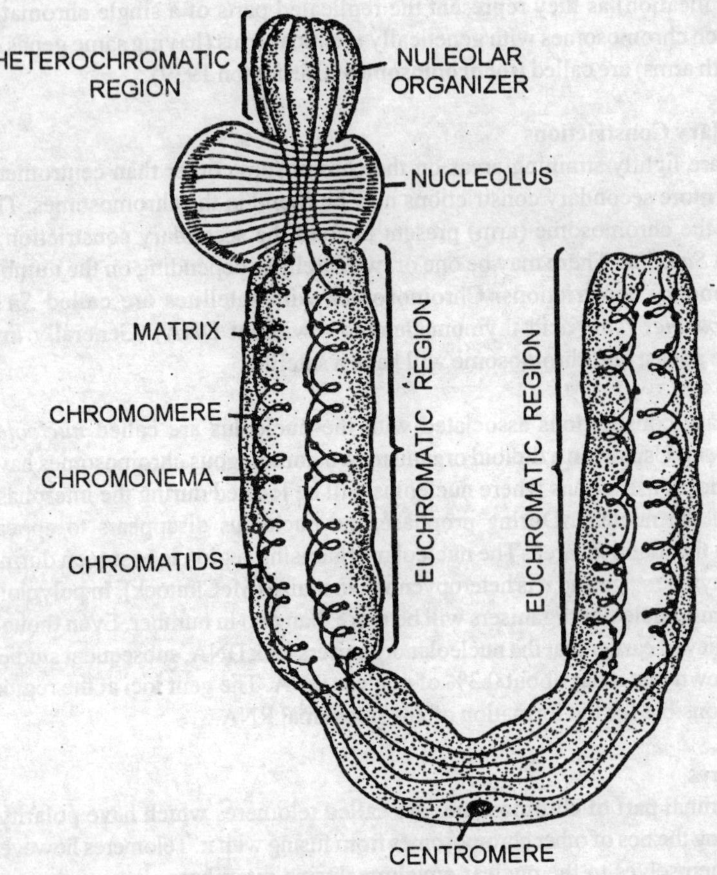

Fig. 3.5 Chromosomes
Structure of a typical chromosome showing a nucleolar organiser

Histones are basic proteins associated with eukaryotic nuclear DNA in the ratio of 1:1 in quite a number of plants and animals. They are the main structural proteins of chromosomes.

Based on the amino acid sequence they are categorized into five principal classes H1(H5), H2a, H2b, H3 and H4.

Histones are characterized by the total lack of *tryptophan*, but have an abundance of *arginine* and *lysine.* Other modifications of histone proteins areacetylation, methylation and phosphorylation.

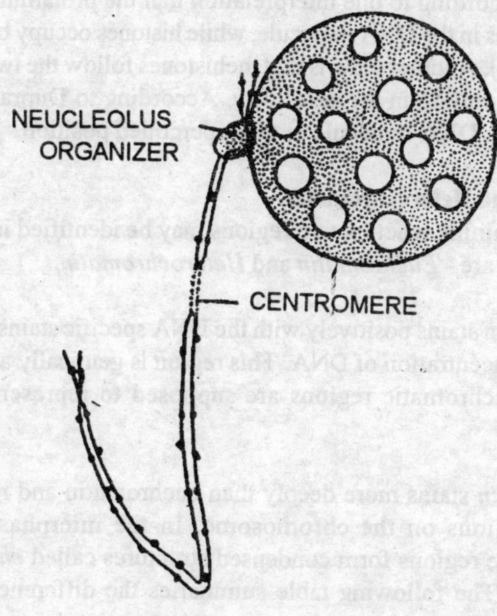

Fig. 3.6 Chromosomes
Nucleolar organiser (enlarged)

Functions of Histones

They act as gene repressors; but this may be a general activity rather than being specific as most of the histones have identical composition. Histones perhaps provide a structural framework for encompassing the DNA molecule.

Synthesis of Histones

In the interphase nucleus, histones are synthesised during the 'S' phase. Some histones may also be synthesized at other stages. The eukaryotic DNA has certain genes called repetitive genes coding for histones.

Non histone proteins

These are acidic proteins with a very high molecular weight when compared with that of histones. NH proteins are known to initiate genetic activity, while histones are known to depress genetic activity.

DNA, Protein Association in the Chromosome

There are different views regarding the association of DNA with the proteins. It is possible according to one interpretation that the protamines occupy only the small grooves in the DNA molecule, while histones occupy both the grooves (small and large). Another view is that thehistones follow the two helices of the DNA molecule coiling in the same way. According to Dupraw, the histones hold together the DNA molecule in the supercoiled position.

Euchromatin and Heterochromatin

Based on the staining reaction two regions may be identified in the chromatin material. These are - *Euchromatin* and *Heterochromatin.*

The euchromatin stains positively with the DNA specific stains (basic fuchsin) indicating a concentration of DNA. This region is gentically active and stains lightly. The euchromatic regions are supposed to represent areas of less condensation.

Heterochromatin stains more deeply than euchromatin and represent highly condensed regions on the chromosome. In the interphase nucleus, the heterochromatic regions form condensed structures called *chromocentres* or false nucleoli. The following table summaries the differences between the euchromatin and heterochromatin.

	Heterochromatin	**Euchromatin**
1. Staining	Stains deeply	Stains less deeply
2. Coiling	Found in the region of more coiling and condensation.	Found in the region of less coiling and condensation
3. Replication	Late replication replication in late 'S' phase.	Early replication in early 'S' phase.
4. Acetylation	Does not occur	Occurs.
5. Effect of temperature	Pronounced	Not affected easily.
6. Genetic effect	Inert genetically but few genes are present	Genetically active. More number of genes present.
7. Cross over frequency	Less	Frequent.

Types of Heterochromatin

There are two main types viz., *constitutive heterochromatin* and *adaptive heterochromatin.* The first type shows dark staining reaction (Heteropynosis) always and in all cells, while the second type shows heteropycnosis only in certain cells or only at certain stages in the cycle. Constitutive heterochromatin was originally named satellite DNA (S-DNA). It is mostly inactive during protein synthesis and is repetetive from thepoint of view of base sequence. Adaptive heterochromatin is metabolically inert and comprises of about 2.5% of genome.

Heteropycnosis

This is the staining phenomenon seen in the chromosomes with reference to heterochromatic and euchromatic regions. The heterochromatic regions which stain deeply are said to exhibit heteropycnosis as against euchromatin which stain lightly. However it should be noted that the heterochromatin always does not stain deeply with reference to euchromatin. While the euchromatic region is always constant in its behaviour (non heterochromatin) towards staining (has constant coils), the heterochromatic region may stain deeply or lightly and this is known as *differential heteropycnosis.* Differential heteropycnosis is of two types - positive and negative.

The areas of heterochromatin showing positive heteropycnosis are more condensed and deeper staining than euchromatic region. In areas of positive heteropcnosis there is more of nucleic acids and proteins. Chromosomes with positive heteropycnosis in prophase appear like metaphase chromosomes.

The areas of heterochromatin showing negative heteropycnosis are possibly less condensed and stain lighter than euchromatic region. Chromosomes with negative heteropycnosis have a hazy outline and metaphase chromosomes appear like prophase chromosomes.

CHROMOSOME MODELS

The chromosome consists of DNA and proteins, but the way in which the two are arranged in the chromosome is a matter of debate. The main aspects of the chromosome structure are (a) the number of strand or strands of DNA (b) whether the DNA molecule is continuous throughout the length of the chromosome or whether it is interrupted and (c) what is the mode of association of protein and DNA.

Various chromosome models have been proposed and these fall into two basic categories viz., *Multiple strand models* and *single strand models.*

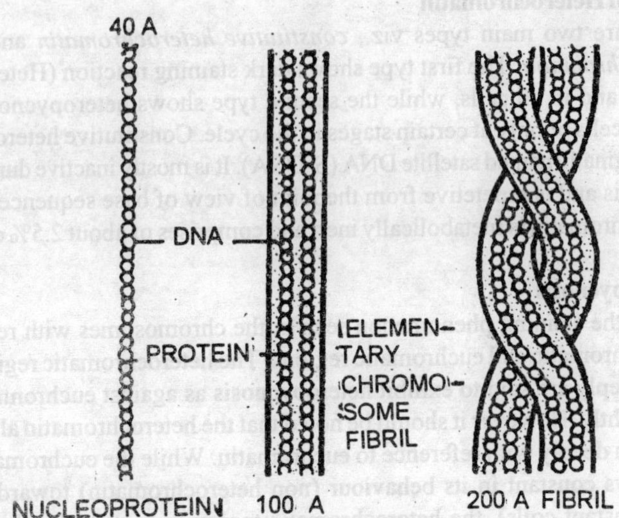

Fig. 3.7 Chromosomes
Ris' multistrand model

Multiple strand models

According to this, the chromosome consists of several DNA and protein strands. The work of Nebel (1938) on radiation effect on chromosomes have shown that there are at least four sub units. Similarly some models suggest the existence of 4,8,16,32 or even 64 fibrils with each fibril having about 20-40 DNA molecules. The following are some of the multiple strand models :-

Simple multistrand model

Accoiding to this model proposed by Steffensen (1961), the chromosome has 64 molecules of DNA arranged parallely and twisted around each other. If one agrees with this model, the replication of DNA would be a difficult process.

Ris' Multistrand Model

In this model, the DNA and histone form a complex 40 Å wide fibrlil Two such fubrils according to Ris (1961) combine to form an *elementary chromosome fibril,* two of these again combine to form a fibril which is 200 Å wide. These fibrils form the chromatids. Each chromatid has 16n to 32 elementary chromosome fibrils. But there isno experimental proof for this work. In fact Ris himselv rejected this model later.

Single Strand Models

Most of the current evidence available now indicates that the chromosomes (barring the abnormal chromosomes) are made up of a single strand of DNA. Experimental evidence by X-ray radiation have suggested the whole of the chromosome is composed of one single coiled molecule of DNA. The following are some of the single strand models.

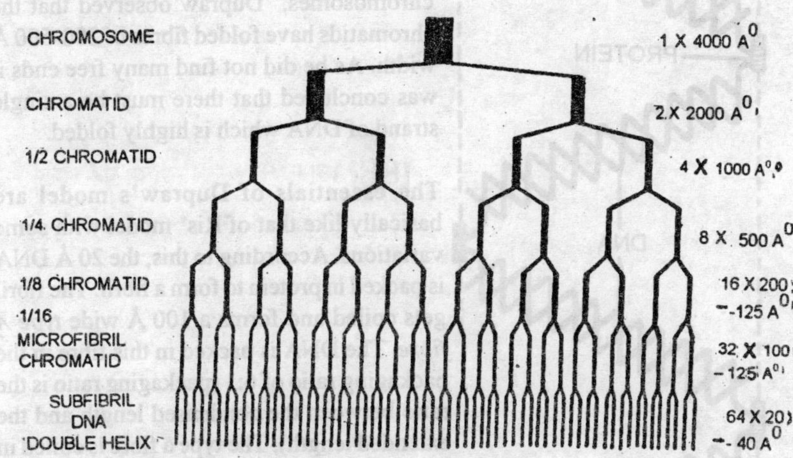

Fig. 3.8 Chromosomes
Arrangement within a chromosome based on multistrand model

The Freese - Taylore Model

According to this model originally proposed by Freese (1958) and later modified by Taylor (1959), there are two chains of proteins with the DNA molecule attaching itself to these like the rungs of a ladder. The proteins linkers maintain the shape of the DNA molecule. The two protein chains get closely appressedin the chromosome resulting in the coiling of DNA.

The original model suggested by Taylor (1957) is called the *side chain model* or *centipede model.* According to this model, the chromosome consists of a central backbone of protein from which the loops of DNA spring up like the legs of a centipede. According to him, replication is possible when the protein backbone made up of two chains, separates longitudinally into two and in theprocess separates also the DNA coils, with each strand having a single strand of DNA. But this is not in conformity with the linear arrangement of the genes in the chromosome.

Nebel's coiled coil model

According to this, the chromosome is made of a single fibril (see his earlier model) is folded to form a 100 Åwide nucleo - histone fibril. The nucleohistone

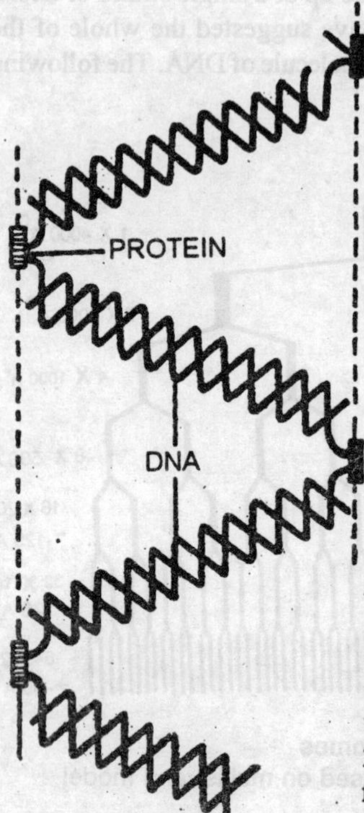

Fig. 3.9 Chromosomes
The Freese-Taylor model

fibril folds because of calcium bridges to form a 250 Å wide *basic fibril.* Further folding of the basic fibril forms the chromosome.

Folded fibre model of Dupraw (1966)

While studying leucocyte (human) chromosomes, Dupraw observed that the chromatids have folded fibres of 200-500 Å width. As he did not find many free ends it was concluded that there must be a single strand of DNA which is highly folded.

The essentials of Dupraw's model are basically like that of Ris' model with some variations. According to this, the 20 Å DNA is packed in protein to form a fibril. The fibril gets coiled and forms a 100 Å wide *type A fibre.* The DNA is packed in this fibre in the packaging ratio of 6:1 (packaging ratio is the ratio between the condensed length and the extended length). The type a fibre is coiled in the ratio of 10 : 1 to form type B fibre which is 200-250 Å wide. The type B fibre is extensively folded to form the chromatid. This has been supported by many workers (McDermot 1968, Fedric 1969, Lampert 1969 etc). Though many scientists have supported Dupraw's view, White, M.J.D. (1973) strikes a discordant note when he does not agree with the folded fibre model and he considers it extraordinary that the DNA of all the chromosomes are linked in the form of a ring. Dupraw's model is now regarded as untenable in view of the fact that the DNA itself forms a wrap around the histones to form *nucleosomes.*

Nucleosome Model

This model proposed by Woodcock (1973) suggests that the chromatid is a string of beads made up of repeating units. These beads called nucleosomes are oblate particles with a diametere of 100 Å. They have a core of histones 40 Å high and 80 Å wide, around which is found DNA in a super helical structure. Finch and Clug (1976) have proposed an arrangement of chromatin in the chromosome model of nucleosomes. The coiled DNA according to them is in the wtring of nucleosomes. The nucleosomes string is then coiled into 300 Å wide solenoid, which in turn coils to form a *supersolenoid* with a diameter of

4000 Å. This super solenoid structure is the unit fibre (chromonema seen in light microscope) whose coils constitute the chromosomes.

The eact nature of the chromosome arrangement of proteins and nucleic acids may be said to be still an open question.

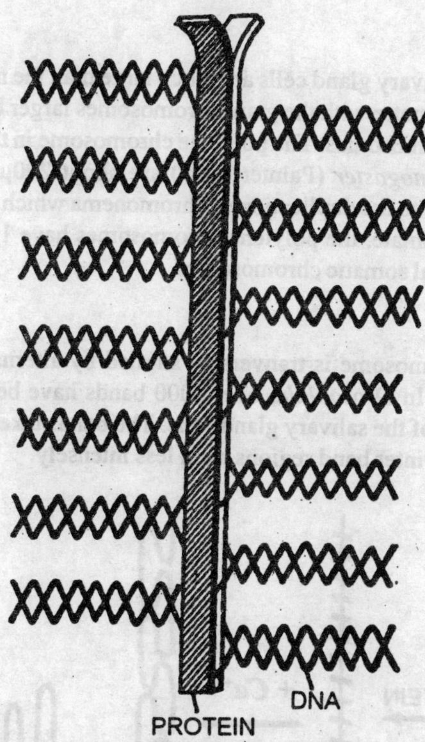

Fig. 3.10 Chromosomes
The centipede model of Taylor

UNUSUAL CHROMOSOMES

The above discussion on chromosomes pertains to normal ones. There are however abnormal or giant chromosomes encountered in many instances. These are the *polytene chromosomes* in the salivary glands of fruit flies, the *lamp brush chromosomes* in the oocytes ofamphibians, and the *supernumerary* or *B chromosomes.*

Polytene Chromosomes

These multithreaded chromosomes were first reported by Balbiani in 1881 while observing the suqash of salivary cells of *Chironomous*. Subsequently they have been seen in the larvae of dipterans (*Drosophila* etc). The name polytene was suggested by Kollar due to the apperance of many threads. Besides salivary glands the polytene chromosomes also occur in rectal epithelium and malphigian tubules.

The nuclei in the salivary gland cells are much large than the normal ones and are about 25μ in diametre with the giant chromosomes larger by 50-200 times than the normal chromosomes. The polytene chromosome in the salivary cells of *Drosophila melanogaster* (Painter 1933) are about 200μ in length. The enormous size is due to the duplication of chromonema which do not separate. According to an estimate, the polytene chromosomes have 1000 times more DNA than the normal somatic chromosomes.

Structure

Each polytene chromosome is tranversely striated by alternating bands and *inter band regions*. In *Drosophila*, over 5000 bands have been found in the four chromosomes of the salivary gland cells. The bands take an intense stain (Feulgen) while the inter band regions stain less intensely.

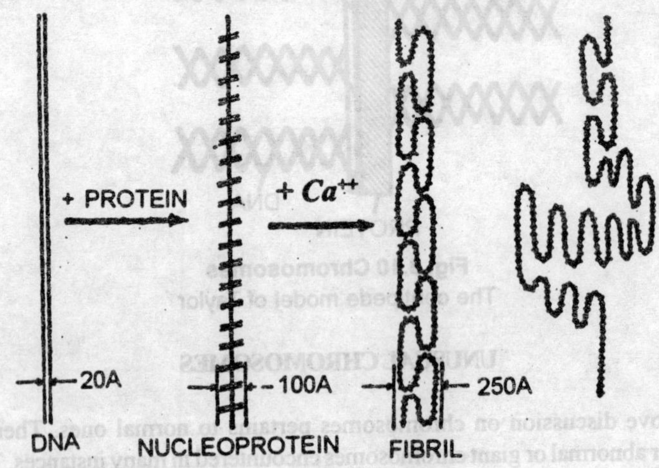

Fig. 3.11 Chromosomes
Ris' modified model

The bands have chromomeres arranged at right angles to the long axis of the chromosome. The bands actually represent regions of tight coils of chromonema. Van Herwerden (1911) interpreted these bands as gyres of a continuous spirl. This view was later supported by Kostoff (1930). The bands have ahigh DNA content and absorb U.V. light at 2600 Å.

The interbands stain lightly and they are negative to feulgen stain indicating a low concentration of DNA. The chromonema in the interband region are less coiled and absorb very little U.V. light.

Painter (1933) and Bridges (1936) have shown that in *Drosophila* the bands are clearly associated with the genes. Association of certain defects due to mutated genes have also been associated with the bands. Certain recent studies however have shown that together with the bands, the interband regions also have genes.

Chromosome puffs

These are certain interesting structures associated with the bands in the giant chromosomes called *chromosome puffs* or *Balbiani rings.* The swelling in bands at certain stages are given the name chromosome puffs. It is quite possible that these puffs are associated with the metabolic activities and represent areas of active RNA synthesis. According to Beerman and Bahr (1954), the coils in the puff region open out to form many loops. The larger chromosome puffs are called *Balbiani rings* and they show a high DNA and mRNA content.

Lamp Brush Chromosomes

The giant chromosomes found in the oocytes of fish, amphibians, reptiles and aves look like the brush used to clean the chimney of a kerosene lamp-hence the name *lamp brush chromosomes.* These are very large and are characterised by loops arising out of the main axis laterally. Discovered first by Flemming in 1882, in amphibian oocytes, they were studied in great detail by J.Rucket (1892) in the oocytes of shark. Rarely lamp brush chromosomes have also been found in some invertebrates (Sepia, Echinaster, insects etc) but they are not as large as those of vertebrates. It seems they are present in certain plants also (Grun 1958).

In somee organisms, lamp brush chromosomes as long as 1000 microns and 20 microns wide have been reported during early prophase I. Towards the end of prophase, they get reduced in size. There are reports of lamp brush chromosomes as long as 5900 microns in some salamander oocytes. These chromosomes are elastic and can be stretched.

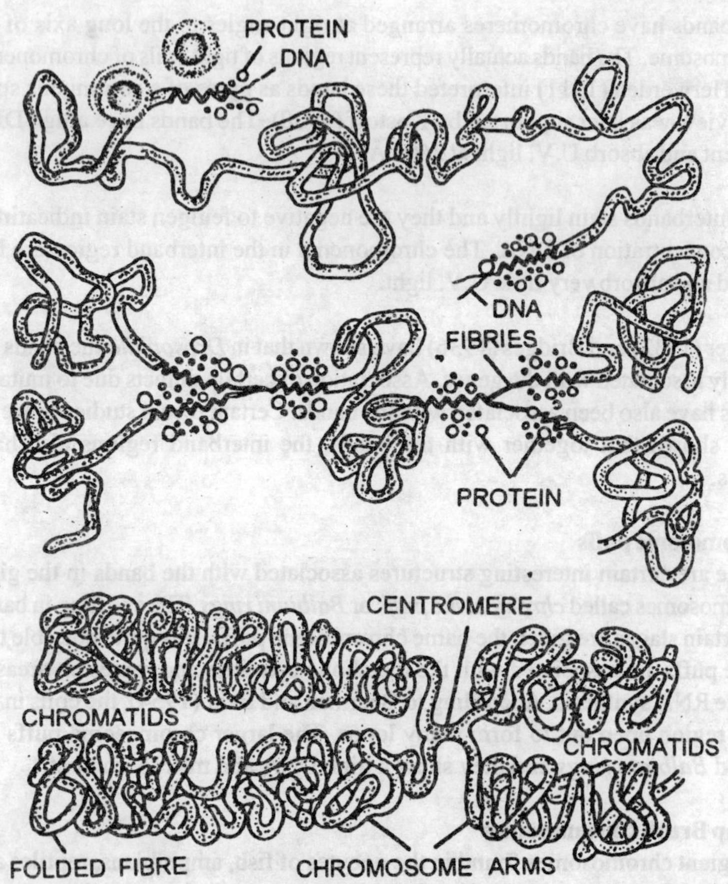

Fig. 3.12 Chromosomes
Folded fibre model of Dupraw

Structure

There are two main parts viz., the main axis and lateral loops. The main axis has two chromatids (the homologues have four) and is rich in DNA. The axis is in continuity with the loop axis. The tips of theaxis are without loops and are the regions of telomeres. The centromeric region is also without loops. At certain regions along their length the chromonema of the axis get coiled and form the chromomeres. The chromomeres are found in pairs. Electron microscopic studies of the lamp brush chromosomes of the salamander *Triturus* (Miller and Beatty 1969) have shown the axial fibre containing dense granules which are probably the molecules of the enzyme RNA *polymerase*.

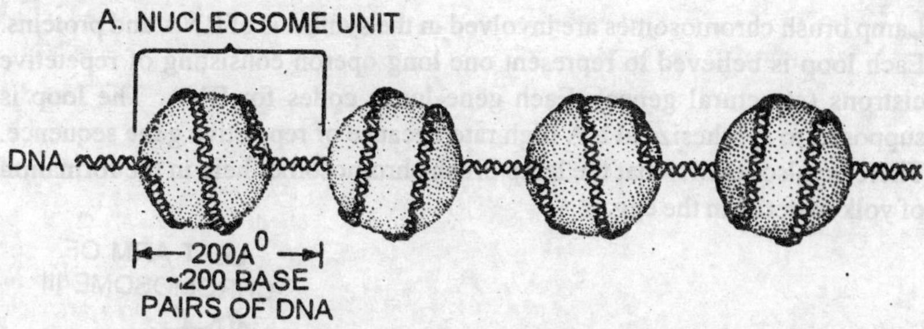

Fig. 3.13 Chromosomes
Nucleosome model

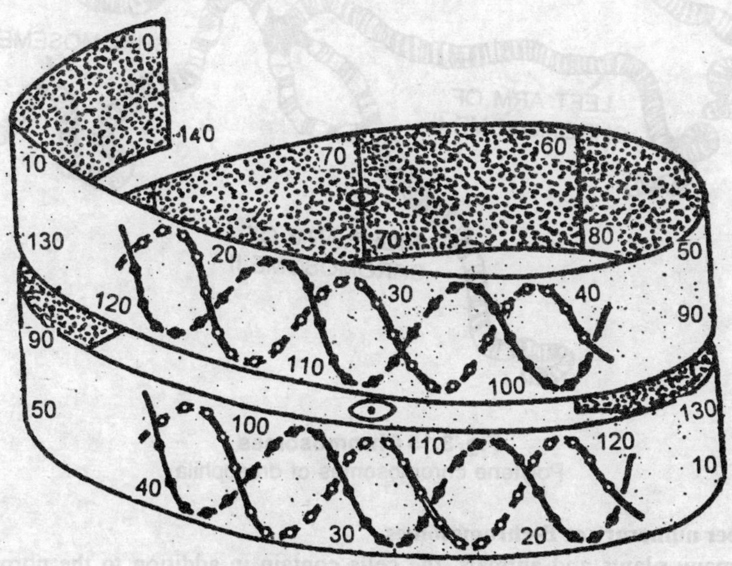

Fig. 3.14 Chromosomes
Schematic representation of DNA super helix in the nucleosome

The loops are of two main types viz., *typical* and *special*. Most of the loops which are typical consist of a bent central axis from which are given out RNA fibrils. These RNA fibrils are of progressively increasing length as a result one side of theloop is pronouncedly thick. The special loops are symmetrical and have granules at the end of RNA fibrils. Each loop axis has an axial fibre covered with a matrix. The diametre of the loop axis varies from 30-50 Å.
The axiaL fibre is made up of DNA< while the matrix has RNA and proteins.

Lamp brush chromosomes are involved in the synthesis of RNA and proteins. Each loop is believed to represent one long operon consisting of repetetive cistrons (structural genes). Each gene locus codes for RNA. The loop is supposed to synthesize at a A high rate because of repetetive gene sequence. There arealso reports that the lamp brush chromosomes help in the formation of yolk material in the egg.

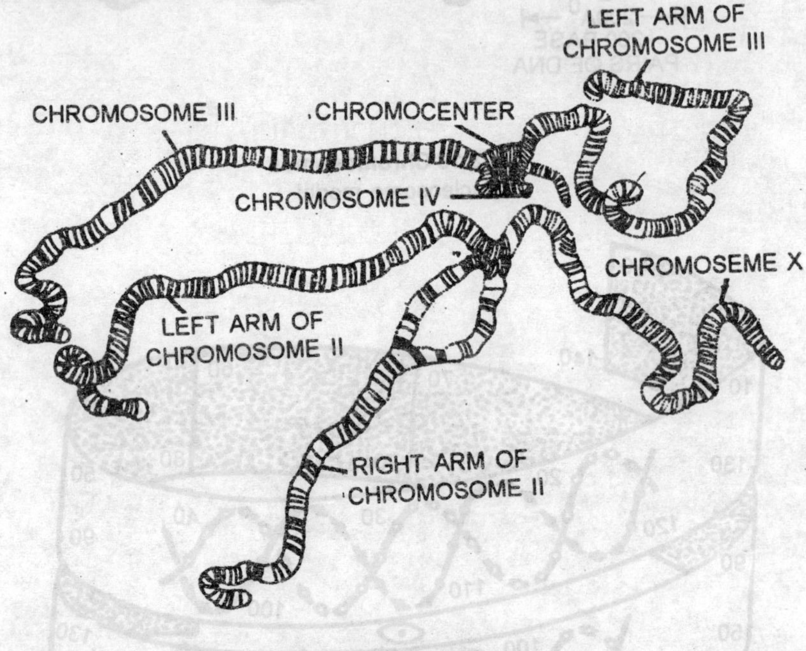

Fig. 3.15 Chromosomes
Polytene chromosomes of drosophila

Super numerary or B-chromosomes

In many plants and animals, the cells contain in addition to the normal complement, one or more extra chromosomes called supernumerary or accessory or B-chromosomes. These were first discovered in Hemipteran insects (Wilson 1905); since then they have been found in a number of insects and plants.

The supernumerary chromosomes are generally smaller than the others and are genetically inert hence produce little or no phenotypic expression. This suggest that these chromosomes may largely be heterochromatic which is supported by their differential staining activity. But this is not universal, for in *Tradescantia* they are entirely euchromatic, while in maize they are part euchromatic and part heterochromatic.

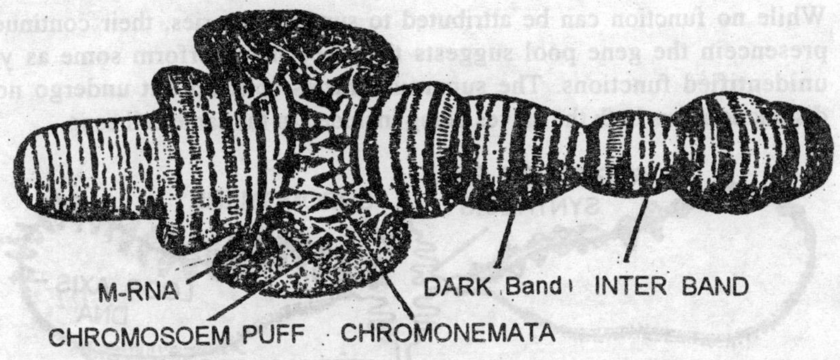

Fig. 3.16 Chromosomes
Polytene chromosomes of a Dipteran insect

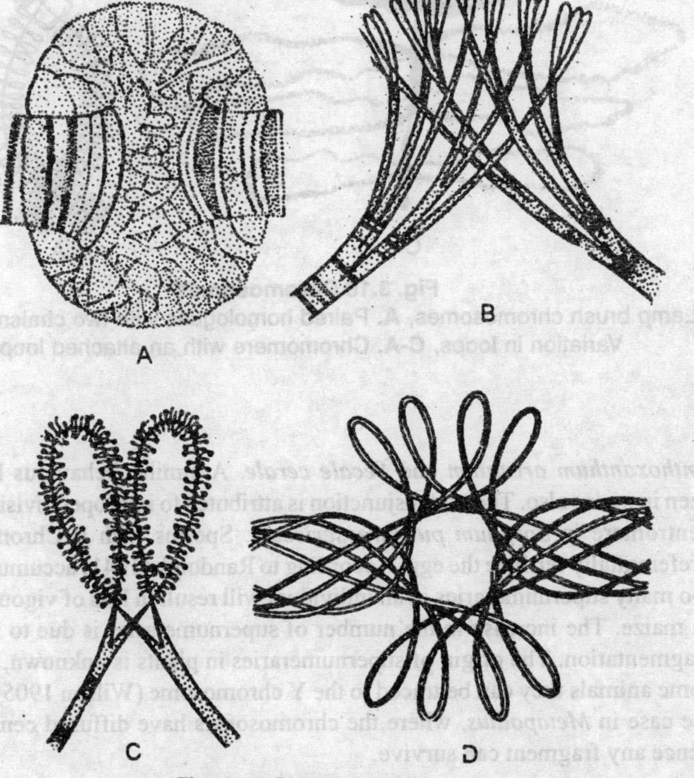

Fig. 3.17 Chromosomes
Balbiani ring in a polytene chromosome

While no function can be attributed to supernumeraries, their continued presencein the gene pool suggests that they may perform some as yet unidentified functions. The supernumeraries split up but undergo non disjunction thus both the halves being included in the same cell as in

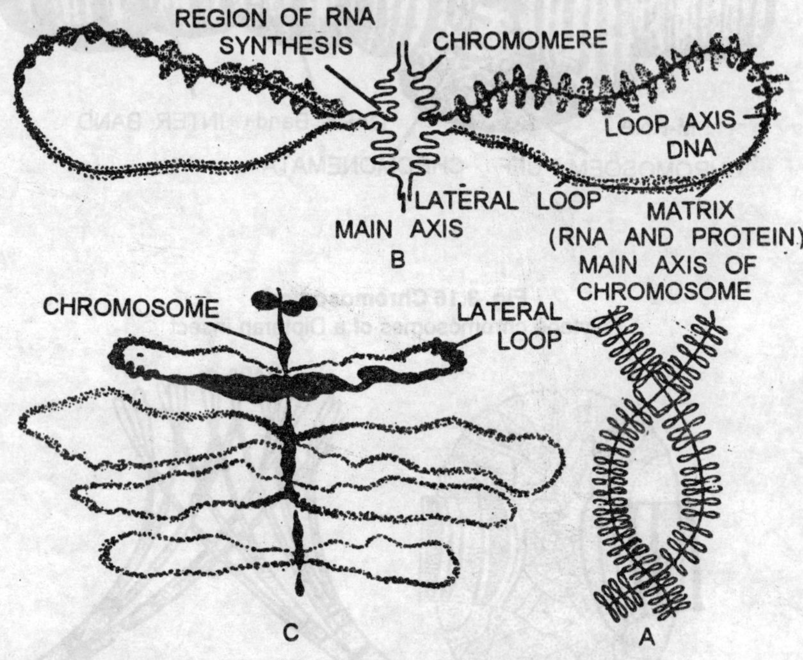

Fig. 3.18 Chromosomes
Lamp brush chromosomes, **A.** Paired homologues with two chaismata, **B.** Variation in loops, **C-A.** Chromomere with an attached loop.

Anthoxanthum aristatum and *Secale cerale.* A similar behavious has been seen in maize also. The non disjunction is attributed to improper division of the centromere in *Sorghum purpureosericeum.* Sperms with B-Chromosomes preferentially fertilize the egg. According to Randolph (1941) accumulation of too many supernumeraries in an individual will result in loss of vigour as seen in maize. The increase in the number of supernumeraries is due to irregular fragmentation. The origin of supernumeraries in plants is unknown, while in some animals they can be traced to the Y chromosome (Wilson 1905). This is the case in *Metapodius,* where the chromosomes have diffused centromere, hence any fragment can survive.

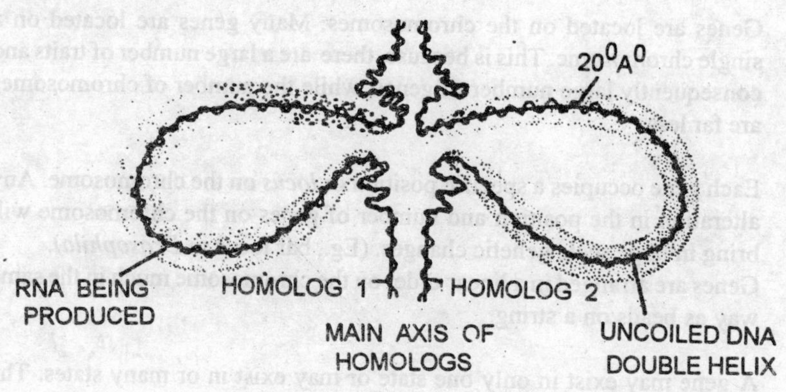

Fig. 3.19 Chromosomes
Lateral loop of a lampbrush chromosome enlarged

GENES

The possibility of the existence of discrete hereditary particles came to be known due to the path breaking work of John Gregor Mendel. It was he who provided some order into the concept of heredity which was otherwise an assemblage of weird theories proposed by many biologists others. Mendel, by his simple but ingenious experiments proved that the heredity is controlled by discrete particles called 'factors' which seggregate, recombine and seggregate again producing a variety of characters. Thus Mendel provided the proof of existence of a physical basis for heredity. The factors of Mendel later came to be known as genes. The concept of gene has undergone a sea change eversince Mendel introduced it, but its essentials remain the same.

The term *gene* was coined by Johannsen in the year 1909 for Mendelian factors. He however knew very little regarding the nature and structure of the gene.

Gene concept

The gene concept was first introduced by Sutton and works of Morghan, Bridges, Boveri etc proved this hypothesis. The essential features of gene concept are summarised below -

1. **Genes decide the physical as well as physiological characters. These are transmitted from parents tooffspring, thus serving as a genetic link between one generation and the other.**

2. Genes are located on the chromosomes. Many genes are located on a single chromosome. This is because, there are a large number of traits and consequently large number of genes, while the number of chromosomes are far less.

3. Each gene occupies a specific position or *locus* on the chromosome. Any alteration in the position and number of genes on the chromosome will bring in large scale genetic changes. (Eg., bar locus in *Drosophila).*
4. Genes are arranged in a linear order on the chromosome much in the same way as beads on a string.

5. A gene may exist in only one state or may exist in or many states. The alternative state of existence of the gene is known as *allele.* Most of the genes are *diallelic* i.e., they exist in two states - *dominant* and *recessive;* of these one is the original and the other is mutant. In some genes, due to series of mutations in the same locus instead of two alleles, there will be a number of alleles. This is known as *multiple alleles* (eg., blood groups in man). They represent a series of less and less dominant genes starting with the wild one.

   In all cases however, even in dialletic series a gene need not be completely dominant or recessive.

6. Genes undergo change, resulting in their altered structure and function. This is known as mutation.
7. Genes duplicate themselves precisely andproduce copies. This is known as replication
8. Genes control the production of proteins (enzymes) and thereby regulate metabolic activity.
9. Chemically genes are made up of segments of DNA. This has been proved by the genetic transformation experiments of Avery Macleod and McCarty (1944). In some plant viruses however genes are composed of only RNA (eg., TMV).

Modern Definitions of Gene

Benzer clearly established the relation between various genetic phenomena and the DNA molecule. He coined different terms to denote the various aspects of gene structure and function. These are - *Recon, Muton* and *Cistron.*

Recon

Genetic research has indicated that on the DNA molecule, short segments exist which can be separated from the rest during crossing over. These segments are

refered to as genes. In other words, gene represents a unit of recombination of the DNA molecule. Within a gene crossing over or recombination is not possible. But studies have indicated that within a gene also, recombination is possible. Hence a sub unit of gene which is capable of recombination is called a *recon.* Recombination studies on microbes have indicated that structurally, recon consists of one or two pairs of nucleotides.
Crossing over within a gene was demonstrated by Benzer in T4 bacteriophages.

Cistron
Garrod (1908), thepioneer who established the relationship between gene and its biochemical function argued that the gene directly controls enzyme synthesis. This was later elaborated by Beadle and tatum (1941) as '*one gene one enzyme hypothesis*'. The gene was regarded as a single unit to code for a single enzyme protein.

According to Benzee, the cistron represents a segment of the DNA molecule and consists of a linear sequence of nucleotides to control the production of a protein. In most cases thecistron is equivalent to the gene structurally. In *Escherichia coli,* eachcistron has 1500 base pairs. Some cistrons may be as long as 30,000 base pairs. As the genetic code is a triplet, each gene has 3 times the base pairs as compared to thenumber of aminoacids in the polypeptide chain (it codes).

Researches have however indicated that not all genes code for proteins, the hypothesis one gene one enzyme is no longer tenable.

Complon
This is a term used to replace cistron. A complon is said to be a unit of complimentation. Complimentation is a phenomenon, where in the active groups of enzymes (when they are made up of more than one polypeptide chain) are complimentary to one another.

A few other aspects of gene structure and function are given below.

Discontinuous genes
Researches carried out in 70s indicated that biological informationmay nor be carried by the gene continuously all along its length. A gene is split into several sections separated by non coding segments of DNA. Such genes are called *discontinuous genes* or *split genes* or *mosaic genes.*

In split genes there are two sections - *introns* and *exons.*

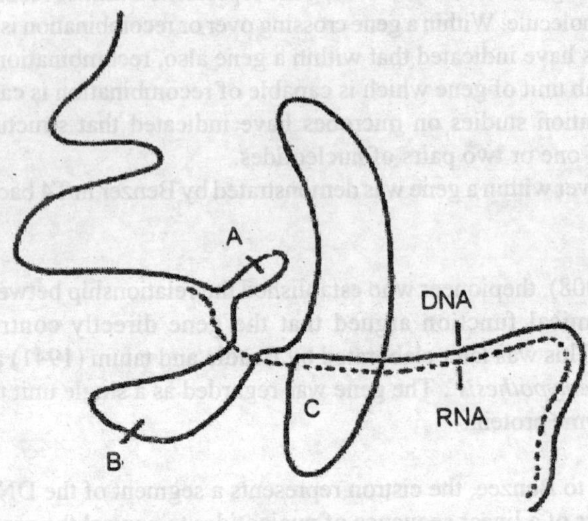

Fig. 3.20 Chromosomes
The hexon gene of Adenovirus. The loops of DNA represent regions of intron

Introns are non coding sections of DNA segment. These are present in genes of eukaryotes and their viruses, but absent in viruses, but absent in prokaryots and their viruses. In eukaryotes mitochondrial genome does not have introns. The mammalian α collagen gene has as many as 52 introns.

Gilbert, has proposed a hypothesis for the origin hypothesis for the origin of new genes by the shuffling of exons between different discontinuous genes.

Transcription from a split gene which has several introns would initially produce a precursoe mRNA which will have several non coding regions. Before translation, the introns have to be removed and the exons reatached to produce a final copy of mRNA with continuous codons. This is called *gene splicing*. Examples of split genes include *Hexons gene* of adenovirus, *ovalubumine gene* of chicken, β globin gene of mice and rabbits. In the ovalbumin gene the intron proteins of mRNA loopout and then these loops are excised.

Overalpping genes

According to the one gene one protein hypothesis of Beadle and Tatum (1940), each gene will separately code for one protein and that genes do not overlap.

But the work of Barrel *et al* (1976) gave the first evidence of overlapping genes in the bacteriophage ϕ X174. They found that two genes code for a particular protein. Usually in organisms the length of DNA should be more than the length of the polypeptide as three nucleotides code for one aminoacid. In phage ϕ X174, however the length of the polypeptide chain is more than that of DNA. This indicates that the proteins coded by two genes is specified by the same of DNA.

Jumping genes

Generally, the gene loci on the chromosome are fixed. But the work of Nobel laureate Barbara Mclintock has shown that genes keep changing their location on the chromosome. The occurrence of jumping genes has also been discovered in bacteria. Streches of DNA in bacteria can move from one location to another, therby altering the expression. The jumping segments of DNA have been called **insertion elements.**

Barbara Mclintock, working on maize has shown the movement of certain streches of DNA known as controlling elements from one location to another.

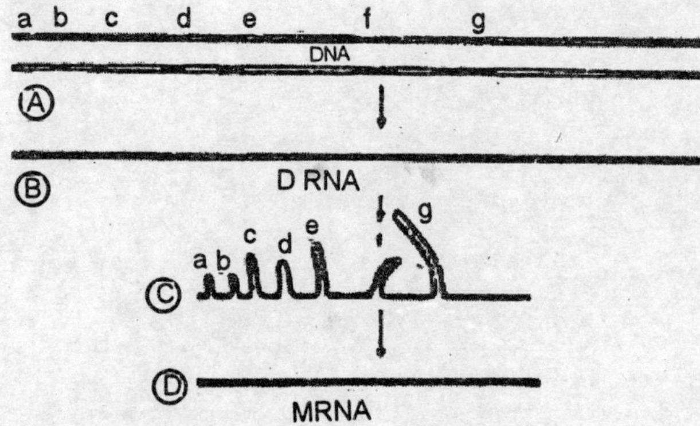

Fig. 3. 21 Chromosomes

Transcription and mRNA in ovalbumin. gene-Differentstages.(In A and B blank representsnon coding regions while black represents coding regions)

Pseudogenes

In multicellular organisms, a wide variety of DNA sequences are found which apparently have no function. Some of these sequences are defective copies of functional genes and are therefore called pseudogenes. Pseudogenes have been reported from mouse, *Drosophila* and human beings. A typical example of human pseudogene is α globin and β globin pseudogene.

GENOME IN PROKARYOTES

The hereditary material in all organisms is remarkably similar. Having been evolved in the remote past, life has been able to maintain itself because of the ability of the hereditary materials to maintain a basic uniformity in structure and function. The hereditary material in all living beings (whether it is a border line entity like a virus, or an evolved human being) is Nucleic acids without exception, even though, there may be some stray instances of proteins (as prions) believed to carry out inheritance. The structure of the nucleic acids is unique; it has a basic uniformity, yet sufficient plasticity to allow for its manifestation from most primitive to the advanced organism.

In eukaryotic organisms, the genetic material is organised into large entities called chromosomes which allow for proper replication and distribution of genetic material during cell division and reproduction.

In prokaryotic organisms also,the nuclei acids constitute the chromosome, but the chromosome of prokaryotes is a naked strand of DNA (not associated with proteins) lying free in cytoplasm. The entire genetic system of each species is encoded in a single chromosome. Even in viruses, the genome is either DNA or RNA.

The prokaryotic genome is refered to as the chromosome, centrosome, nucleoid, chromatin body or the gneophore. Some traditional cytogeneticists object to the word chromosome and prefer to use the term genophore or nucleoid. However, as Swanson (1981), points out "this is more a matter of semantics than of cytogenetics, since the prokaryotic chromosome not only has the same fundamental structure based on nucleic acid sequences, but it also carries out, the same three basic functions of its eukaryotic counter part. That is, it is the organelle which stores, replicates and transmits the coded information of biological inheritance; which is so regulated in its action that its primary products appear at appropriate times and in appropriate amounts during the life-cycle of the organism; and which is capable of recombining its genetic content with other similar individuals in such a manner that the seggregating progeny of a hybrid union can have different genetic properties". Swanson (1981), therefore, opines that the term "chromosome" can be applied to both eukaryotic and prokaryotic species with distnctions in structure being made at appropriate times.

The chromosome of *Escherichia coli,* and *Salmonella typhimurium,* are probably the most thoroughly mapped genomes of any cellular organism. Hundreds of identifiable genes are located in a single circular linkage group. The entire genome of a bacterium is made up of a single chromosome. Occasionally, a single bacterial cell may have more than one chromosome at a given time. This should not be taken as the cell having two chromosomes. This represents a stage wherein the chromosome has replicated and formed two units, but the cell is yet to divide.

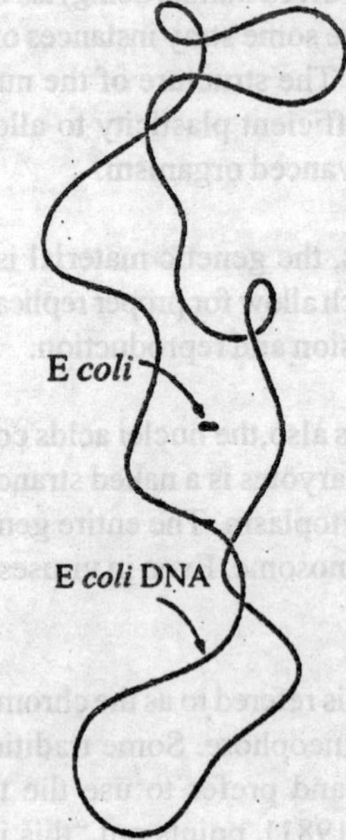

Fig. 4.1 Genome in Prokaryotes
Schematic diagram showing the relative sizes of E.coli and its DNA molecule

The chromosome of *E.coli* is a single, large double helix which is circular in outline. The DNA molecule is supercolied. It (DNA) has no association of histones as in eykaryotic chromosomes. However, small amounts of basic proteins have been found to be associated with the isolated DNA. It is quite likely that these might prevent some sections of the chromosome from transcription thus functioning like repressor proteins. Thus, it can be argued that the absence of DNA-histone association may not be such a fundamental point of difference between prokaryotic and eukaryotic DNA.

In *E.coli,* the total length of the circular DNA is 1300μm, whereas cylindrical bacteria have a maximum diameter of 1μm only. This clearly indicates that the chromosome must be highly folded in the cell.

When the DNA of *E.coli* is isolated by techniques that avoid both DNA breakage and protein denaturation, a highly compact structure called ***nucleoid*** can be seen. This structure contains a DNA molecule, a fixed amount of protein and variable amounts of RNA. As has been pointed out already, the proteins might play a part in repression of transcription.

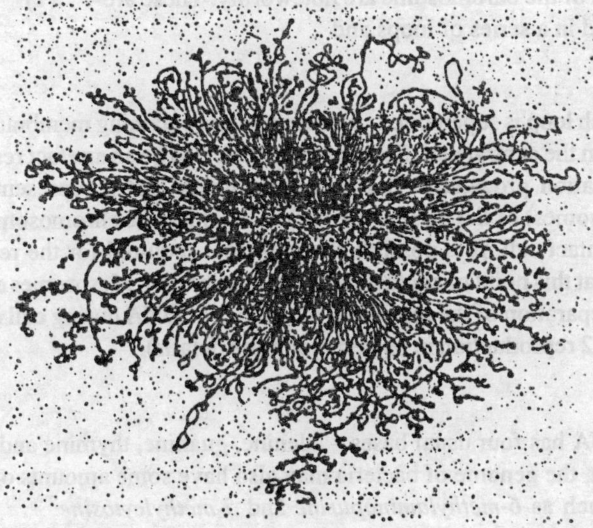

Fig. 4.2 Genome in Prokaryotes
An electron micrograph of an E.coli chromosome showing the multiple loops merging from a central region

The DNA molecule is about 1000 to 1100μm in length with a diameter of 20 Å to 40 Å. Such a molecule has more than 3.235 x 10^6 nucleotide pairs (each pairs is 3.4 Å long and there are 2941 base pairs per micrometer). The molecule has a molecular weight of 2.5 x 10^9 daltons (one dalton is equivalent to the weight of one hydrogen atom). This entire molecule is limited in the nuclear area of the cell which is no more than one micron and has no limiting boundary (such as a nuclear envelop), hence the molecule must be enormously folded.

It has been pointed above, that a small amount of RNA and basic proteins are associated with the DNA molecule. But the fact that the action of RNAse or protease does not alter the basic structure of DNA, indicates that neither RNA nor proteins are necessary for the basic circular natrue of the chromosome.

Detailed Structure of the Chromosome

The chromosome (DNA) of *E.coli* on isolation appears to be a highly folded compact structure. These folds appear to be highly regular and not random. If they were to be random, replication and proper seggregation would not be possible. The chromosome obviously has a high level of organization. Two features of the chromosome are noteworthy. These are – (1) The DNA is arranged in a series of loops and

(2) Each loop is supercoiled. A dense region containing membranous material is seen in the central part. Some people think it to be an artefact resulting out of the isolation procedure. But it is quite likely that it represents fragments (measosomes) of the plasma membrane to which the chromosome is attached in the intact cell. It is possible that enzymes required for the replication are located at this point of contact (with the membrane). Mesosomes do play a role in the separation of the replicated molecules of DNA during cell division. See chapter 2 reproduction in bacteria for more details).

The DNA has four usual bases – adenine, guanine, thymine and cytosine. In addition, the genome of bacteria may also have some amounts of methylated bases such as *6-methylaminopurine* and *5-methylcytosine.*

Introduction of a single strand break into the supercoil by a DNAse enzyme results in an abrupt change from supercoiled to non-supercoiled condition. This is because of free rotation about the opposing sugar phosphate bond. If the nucleoid, however, is nicked (cut), it does not become an open circle. But with each nick it passes through 40–50 intermediate stages in which the supercoiled loops turn into open loops one by one. This is a clear pointer to the fact that a single strand break does not cause free rotation of the entire molecule. The DNA is assumed to be fixed to the proteins in such a way that does not allow overlapping of the individual loops.

The enzyme (DNA gyrase) that brings about DNA replication is also responsible for supercoiling. This can be proved by the addition of *coumermycin,* an inhibitor of DNA gyrase, to the culture medium containing *E.coli* cells. The chromosomes of *E.coli* quickly lose supercoiling (on addition of coumerycin). In addition to DNAgyrase, another enzyme *topoisomerase* I, has the ability of unwinding the supercoils. Hence, it is believed that the two enzymes *–DNA gyrase* and *topoisomerase.* I act in a coordinated fashion and regulate the degree coiling. *In vitro* experiments have shown that purified topoisomerase I enzyme relases supercoiling to a non-supercoiled convalent circle. Mutants of *E.coli* lacking *topiosomerase* I have been seen to possess increased supercoiling to the extent of 32 percent more than the normal.

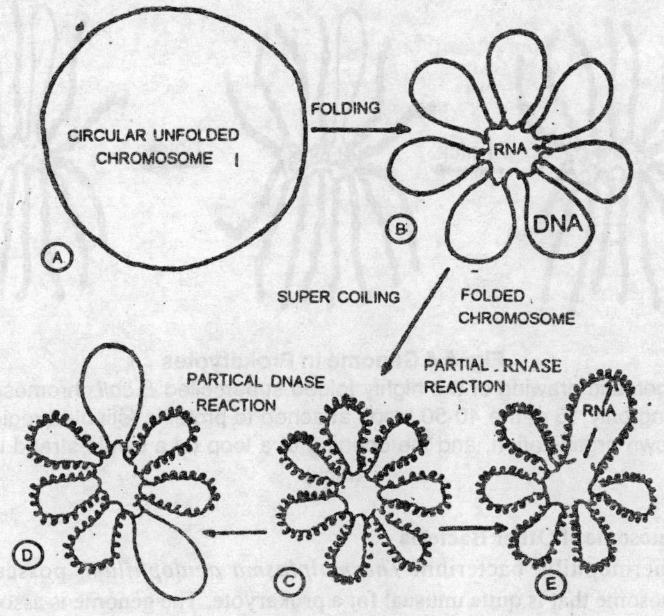

Fig. 4.3 Genome in Prokaryotes
Model of bacterial chromosome

A. Circular unfolded chromosome
B. Folded chromosome
C. Folded and supercoiled chromosome
D. Supercoiling relaxed into loops by DNAse nicks
E. RNAse cuts one RNA molecule

Some interesting observations have been made on these mutants lacking in topoisomerase. I. In *in vitro* conditions, the mutants undergo a further secondary mutation in or near the gyrase gene which seems to reduce the gyrase activity and return the supercoiling to near normal state.

Other Associated Structures of Chromosomes

The isolated genome of *E.coli* has about 30 percent by weight RNA and 10 percent protein. The RNA is largely a newly transcribed mRNA, tRNA or ribosomal RNA which is yet to be released into the cell machinery. A small amount of additional RNA in this is responsible for the folded character of the genome. RNAse activity can partially unfold the genome.

Most of the proteins have been identified as RNA polymerase engaged in transcription. This is quite understandable as a metabolically active bacterial cell would be virtually transcribing throughout its life cycle. A small portion of the protein represents the mesosome.

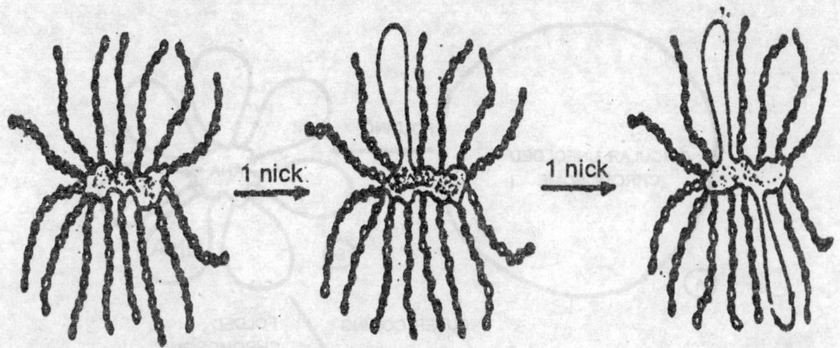

Fig. 4.4 Genome in Prokaryotes

A schematic drawing of the highly folded supercoiled *E.coli* chromosome, showing only 15 of the 40-50 loops attached to proteins (stippled region) of unknown organization, and the opening of a loop by a single-strand break (nick)

Chromosome of Other Bacteria

The thermophilic bacterium *Thermolplasma acidophilum*, possesses a chromosome that is quite unusual for a prokaryote. The genome is associated with a histone like protein which can complex with about 20% of the chromosome. The purpose of such an association is to maintain the stability of DNA during unfavourable (denaturing) environmental conditions.

The chromosome of *Bacillus subtilis,* is circular and has approximately the same dimensions and molecular weight like that of *E.coli.* There are suggestions that the genome of *B.subtils* replicates as a linear molecule rather than as a circular molecule. *Streptomyces coelicolor* and *Salmonella typhimurium* are also known to possess circular genome with the latter acomodating 323 identified genes. Species of *M.mycoplasma* have a small circular chromosome of about 265µ in length. *Haemophilus infleuzae,* the whooping cought causing bacterium, has a chromosome as long as 865µm which often appears linear with two free ends. Whether these ends really represent the points of a linear molecule or represent the ruptures of a circle during extraction, it is difficult to say.

PLASMIDS

The bacterial cell has an additional fragment of nucleic acids in addition to its normal chromosome. This additional genetic material was given the name episome (Jacob, Sccaeffer and Woolmann, 1960). Episome can exist either independently or it can integrate itself with the bacterial chromosome. Plasmids, however, are always independent and never integrate with the bacterial chromosome. Plasmids control a number of traits in bacteria such as –toxin production, antibiotic resistance, pilus production etc. Some people regard that episomes and plasmids are synonymus.

The word plasmid was first used by J.Lederberg (1952). However, the term is restricted only to bacterial cells where circular DNA molecules exist independent of the main genome.

Plasmids are circular double stranded molecules of DNA, and exist independently of chromosomal DNA in bacterial cells. They undergo replication during cell division and are carried to both the daughter cells. The plasmid has a right handed superhelical coil during the resting stage. There is one super helical turn for every 400–600 base pairs. The twisted conformation is known as **covalently closed DNA (CCD).** Cleavage results in the conversion of CCD to open circular form.

The following are some of the properties of the plasmids.

1. They are made up of DNA (double stranded).
2. They are capable of independent replication.
3. Some plasmids control bacterial conjugation.
4. Capable of reversible insertion into bacterial chromosome (only some plasmids).
5. They have the ability to get transfered from one bacterial cell to the other

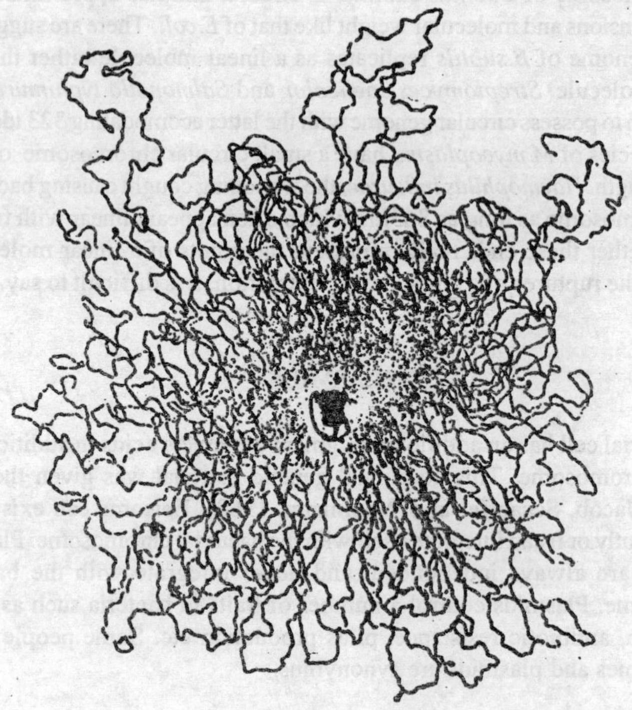

Fig. 4.5 Genome in Prokaryotes
An electron micrograph of the chromosome of the bacterium *Hemophilus infleunzae*

Plasmids are refered to as *conjugative* and *non-conjugative* based on the ability/ inability to promote their transfer during conjugation.

Classification of Plasmids

Basically plasmids are classified into three categories –

1. The F factor or sex factor.
2. The R Plasmid or resistance transfer factor (RTF), and
3. Col factor. There are some plasmids which may share the properties of more than one type of plasmid. Some bacteriologists recognise more than the three types based on some other properties.

F factor

The F factor plasmid confers the ability to the bacterial chromosome to get transferred to another cell (recepient). Hence, the donor is said to be F^+. F factor integrates with the bacterial genome facilitating its transfer. Because of this integration, it is often referred to as an episome.

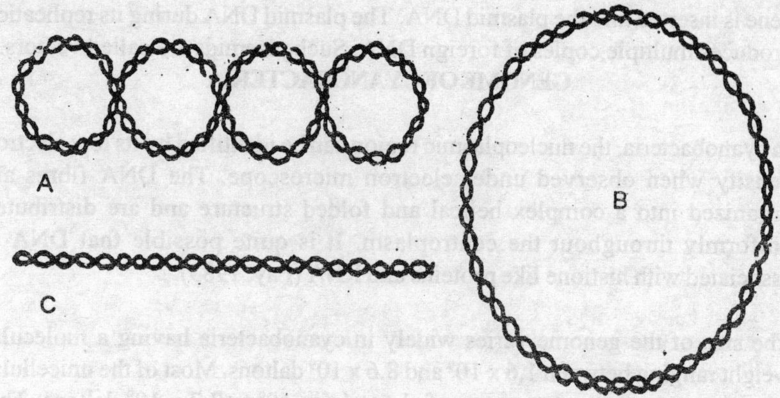

Fig. 4.6 Genome in Prokaryotes
Plasmid DNA, A. Supercoiled form, B. Open nicked circle, C. Linear duplex

R Plasmid

These plasmids, when present, confer on the bacterial cell resistance to several antibiotics such as streptomycin, tetracycline, chloramphenicol, sulphonamid etc. R plasmid mediated antibiotic resistance was first discovered in Japan in 1956, when *Shigella* became resistant to several antibiotics at one step. This was due to the transfer of R plasmid from one strain to the other resulting in not only antibiotic resistance but also the ability of a donor (it can donate R factor to other sensitive strains making them resistant).

R plasmid is a small extrachromosomal DNA ring. Sometimes, the R plasmid has two parts – (a) Part havin drug resistant genes and (b) Part having plasmid transfer genes.

The col factor (Plasmid)

Certain bacteria like *Escherichia coli, Shigella, Salmonella* etc., produce toxins called *colicins.* These are toxic proteins and can kill other bacteria. The ability to produce *colicin* rests in the plasmid called col plasmid or *colicinogenic factor.* Col factors resemble phages since they cause the death of bacteria, but they differ from phages as they are not released out during the lysis of the bacterial cell. Like the F factors, colplasmids also can be transferred from one bacterial cell to the other.

The number of plasmids per bacterial cell varies. It may be *a single copy* or *multiple copies.* Multiple copy plasmids which replicate independently have been used in gene cloning experiments. A plasmid is cleaved and a desired gene is inserted into the plasmid DNA. The plasmid DNA during its replication produces multiple copies of foreign DNA. Such plasmids are called vectors.

GENOME OF CYANOBACTERIA

In cyanobacteria, the nucleoplasmic region can be identified by its low electron density when observed under electron microscope. The DNA fibres are organized into a complex helical and folded structure and are distributed uniformly throughout the centroplasm. It is quite possible that DNA is associated with histione like proteins and RNA (Fay, 1983).

The size of the genome varies widely in cyanobacteria having a molecular weight ranging between 1.6 x 10^9 and 8.6 x 10^9 daltons. Most of the unicellular cyanobacteria possess genomes of about 1.6 x 10^9 to 2.7 x 10^9 daltons. This compares favourably with similar genomes found in other prokaryotes. Some filamentous cyanobacteria, however, have larger genomes. The distribution of genome sizes in cyanobacteria falls into four categories – 2.2, 3.6, 5.0 and 7.4 x 10^9 daltons. Members with a more complex morphology obviously possess larger genomes than those with a simple structure. There is a theory that the modern day prokaryotes evolved from a primitive bacterium with a small genome size of about 1.2 x 10^9 daltons. Replication of this genome during the course of evolution is supposed to have given rise to simple integer multiples seen in modern bacterial genome.

THE VIRAL GENOME

The core of the virion is made up of nucleic acids. A virus has only DNA or RNA, never both together. Within these parameters, however, considerable diversity exists. The nucleic acids may be double stranded, single stranded, linear or circular. Some have plus polarity; others may have minus polarity. Four types of nucleic acids are found in viruses with reference to the number of strands. These are –

1. Single stranded DNA —– (ss DNA) e.g., Colliphage virus
2. Double stranded DNA (ds DNA) e.g., Herpes virus
3. Single stranded RNA (ssRNA) e.g., TMV
4. Double stranded RNA (dsRNA) e.g., Reovirus

Single stranded DNA is found in ϕX174 virus. It was first discovered by Sinscheimer *et al,* in 1950's. ssDNA may be linear (Paroviruses), or circular (ϕX174 virus). The ssDNA becomes double stranded during replication and at that time it is known as the replicative form.

Double stranded DNA is found in a number of animal viruses and bacteriophages. It has a variety of forms – linear (bacteriophages), cross linked (vaccinia virus) or closed circular duplex as in papova Viruses.

Single stranded RNA is found in a variety of animal viruses and icosahedral plant viruses. The strand may be plus (infectious), as in RNA bacteriophages, togaviruses etc. or minus (non infectious), as in rhabdoviruses and paramyxoviruses. Plus ssRNA acts directly as mRNA and translates proteins on the ribosomes of bacteria, whereas minus ssRNA first transcribes an mRNA and then it (mRNA) translates proteins (hence called non-infectious).

Double stranded RNA is found in animal viruses like reovirus, blue tongue virus etc. The dsRNA has 10 or more segments.

The nuclei acid core of Viruses may be summarised as follows –

1. Plant Viruses have only RNA (ss or ds), with the exception of cauliflower mosaic Viruses (DNA virus).
2. Animal Viruses have RNA (ss or ds), and DNA (only ds no ss).
3. Bacteriophages have DNA (ss or ds) or RNA (ss or ds). However, most phages have DNA Viruses.

5

REPLICATION OF DNA (*In Prokaryotes*)

The DNA molecule regulates all the activities in the cell which includes both metabolism as well as reproduction. Because of its unique structure, it is able to control the synthesis of other macromolecules of the cell. At the same time when the cell reproduces, the DNA also should duplicate itself to be able to get distributed equally to the daughter cells. These two unique properties of the DNA molecule can be called *Heterocatalytic* and *Auto catalytic* functions. In the heterocatalytic function, DNA regulates the synthesis of other molecules like proteins. In the autocatalytic function which is perhaps very unique to DNA, it is able to make copies of itself in a precise and accurate manner. It is this ability of DNA that is responsible for the maintenance of life and inheritance of characters from one generation to another generation.

In this chapter, the various aspects of DNA replication in prokaryotes is discussed together with a brief account of the enzyme responsible for replication.

Types of Replication

Three different types of replication have been proposed at the molecular level for DNA. These are –

Semiconservative, Conservative and Dispersive.

1. Semiconservative

In this model of replication, the DNA helix unwinds but there is no rupture of a separated polynucleotide strands. Thus, the primary helices of DNA are not completely changed. According to this mechanism, the unwinding of the original strand would be followed by the synthesis of complementary strands for each of the two separated original polynucleotide strands. This mechanism was proposed by Crick and Watson, the discoverers of the molecular structure of DNA. In this method of replication, after replication of the original molecule, two molecules of DNA are formed. Each one will have one parental strand and one newly synthesised strand. Thus, in the semiconservative process, after replication the molecules of the next generation can be regarded as hybrids. The mechanism is called semiconservative because of the two strands of the parental molecule, only one strand remains in the molecule whereas the other goes to another molecule.

2. Conservative

In this mechanism, the original molecule is completely conserved and thus forms a part of the second generation molecule. Both the strands of the parental helix will synthesise an entirely new molecule, the new molecule synthesized would have both of its strands produced newly. After one generation of application, 50% of the molecules will be parental and 50% newly produced molecules. This is called conservative, because the origiral combination of both the parental strands will remain together and will not be separated.

3. Dispersive

In this method of replication, the parental DNA helix breaks up at several points forming many fragments. Each fragment of the helix would replicate and recombine at random to produce two daughter molecules of DNA. This is called dispersive because the original combination of the strands is completely disturbed as they get fragmented into bits.

Evidence for the Semiconservative Mode of Replication

Of the three types of replication mentioned above, there are overwhelming evidence to support the semiconservative mode of replication. Taylor *et al* (1957), demonstrated by autoradiography that the two chromatids during prohpase have one strand old and one strand new material. This clearly indicates that the replication is semiconservative. The following three experiments have clearly given experimental evidence for the semiconservative mode of replication.

1. Meselson and Sthal's Experiment

Meselson and Stahl (1958), conducted experiments on *E.coli* and demonstrated that the semiconservative mode of replication is correct. In their experiments, they used the following two basic principles.

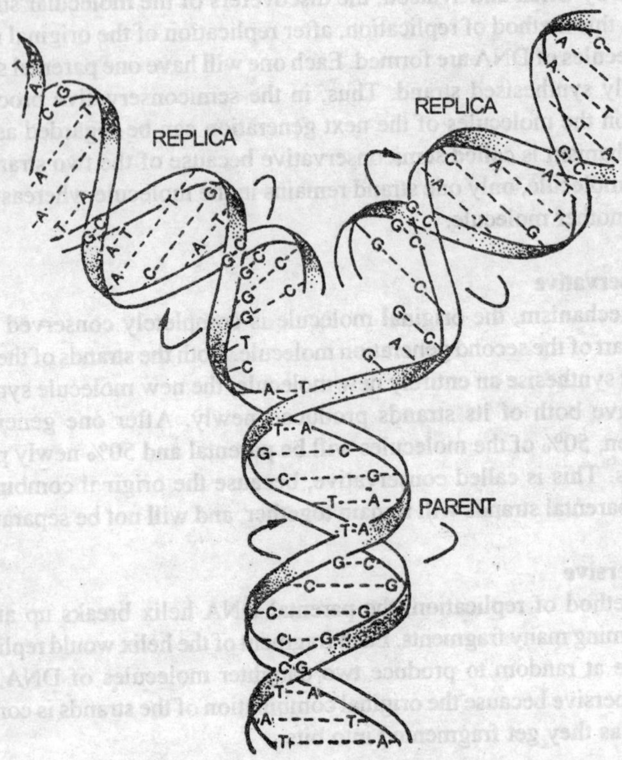

Fig. 5.1 Replication of DNA
Model of replication of DNA

(i) They cultured *E.coli* on a culture medium containing N^{15} (N^{15} is an isotope while normal nitrogen is N^{14}). The bacteria replicated in this medium for a number of generations and as a result, both the strands of the DNA molecule now contain N^{15}. In other words, the DNA of the bacterial populaton was now labelled with heavy nitrogen (N^{15}). The labelled bacteria were then transferred to a culture medium containing N^{14} and are allowed to replicate. After the first generation of replication, of the molecules of DNA, each one possesed one strand with N^{15} and the other strand with N^{14} (this should be the composition of the molecules if the mode of replication is semiconservative). In these molecules, the N^{15} strand represents a parental strand and the lighter N^{14} strand presents a newly synthesised strand. Thus, each molecule can be regarded as a hybrid DNA (having both N^{15} and N^{14}).

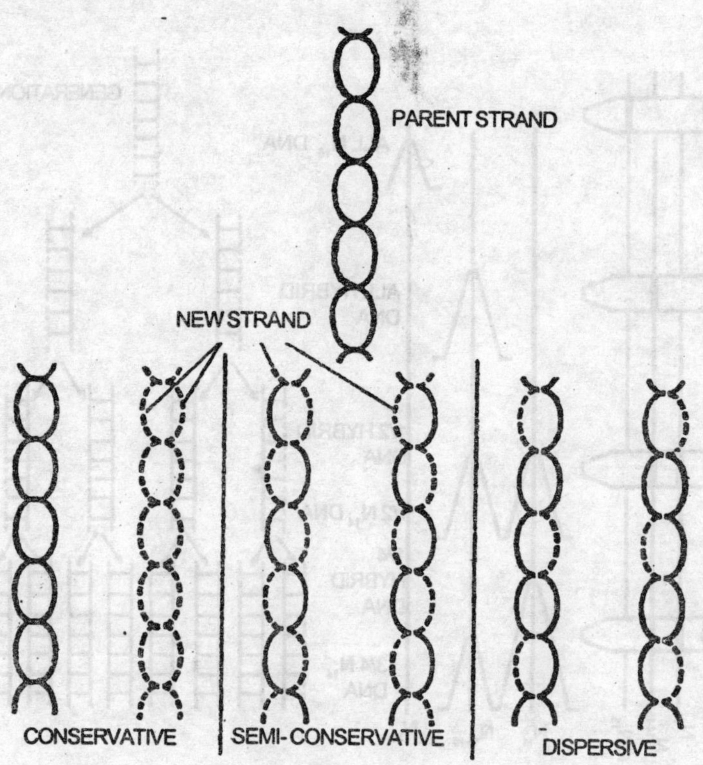

Fig. 5.2 Replication of DNA

Models of DNA replication - Conservative, semiconservative and dispersive

(ii) Meselson and Stahl used the principal of density gradient centrifugation and prepared gradients of Caesium chloride. This density gradient is based on the fact that when heavy salt solutions are subjected to ultra centrifugation, a continuous density gradient is set up (heavy would be at the bottom and lighter ones will be at the top with a range in between in the centrifuge tube). When a substance having a density within the range of this gradient is dissolved in salt solution, the substance will find place at its own level of density. Thus, with this technique even very slight differences in density can be detected.

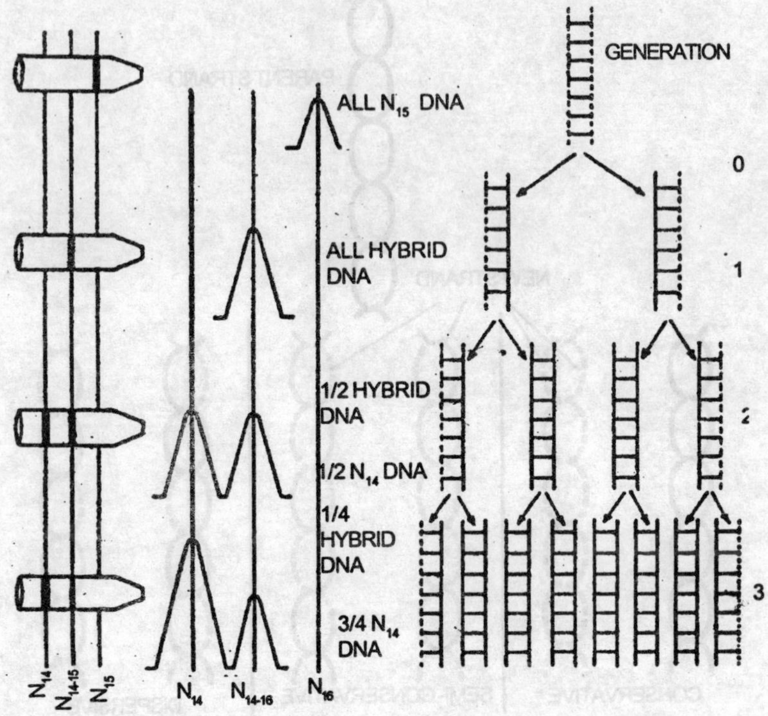

Fig. 5.3 Replication of DNA

Meselson and Stahl experiment in support of the semiconservative model of DNA replication

Based on the principles menthioned above, Messelson and Stahl subjected the bacterial cultures for density gradient centrifugation. At the first instance, N^{15} labelled bacteria were allowed to replicate on a medium containing N^{14}. After the first division, the centifugation showed the hybrid DNA had an intermediate density and formed a band inbetween N^{14} and N^{15} indicating clearly that each molecule contains one N^{14} and N^{15}. In the second generation two bands were found, one having normal N^{14} and the other having a hybrid (N^{14} and N^{15}). This is expected because the labelled nitrogen - N^{15} is getting reduced as the medium has only N^{14} and the original labelled N^{15} gradually gets decreased in the molecules. In the third generation, 3/4ths of the DNA and N^{14} and 1/4th of the DNA was hybrid. If the replication continues for some more generations N^{15} would disappear.

In the above experiment, if we assume that the replication is conservative there

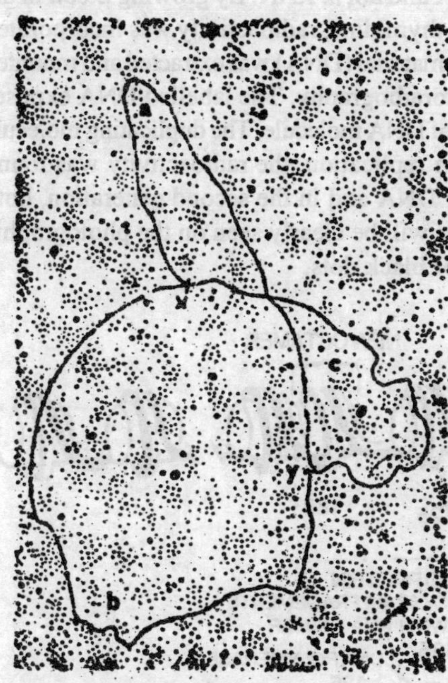

Fig. 5.4 Replication of DNA
Autoradiography of the single circular chromosome of *E. coli* labelled with tritiated thymidine-Cairans experiment

should be two bands in the centrifuge one with N^{15} and the other with N^{14} (newly formed molecule). On the other hand, if the mechanism is dispersive the molecule should spread all over the medium and should form only one band as both N^{14} and N^{15} in different combinations would be there in all the molecules. If the mechanism is semiconservative, the first generation would show only one layer having a mixture of N^{14} and N^{15}, and in the second generation there would, be two layers one N^{14} and the other N^{15} and N^{14}. Gradually, N^{14} would show greater band width and N^{15} would decrease. (Note. In the dispersive mode also only one band is formed in the first generation and it continues to be so for the second generation also. Where as in semiconservative mode, two bands are formed in the second generation).

The above explanation clearly demonstrates that the semiconservative mode of replication proposed by Watson and Crick is the actual method of replication.

2. Cairn's autoradiographic experiment

The semiconservative mode of DNA replication has been proved in the chromosomes of *E.coli* by autoradiographic experiments conducted by Cairns.

Autoradiography is based on the fact that decaying radioactive elements emit electrons which collide with the silver grains of a photographic plate and turn it black. Thus in a autoradiogram, black dots represent positions of decaying radioactive elements. DNA of an organism can be labelled selectively by growing cells in a medium containing radioactive thymidine. Radioactive thymidine is obtained by using H_3– a heavy isotope of hydrogen known as tritium. Cairns used tritiated thymidine as it selectively labels only DNA and not RNA, because thymidine base is found only in DNA and not in RNA. By growing *E.coli* in the culture medium containing tritiated thymidine, radioactivity was incorporated in the daughter DNA molecules. The material (DNA) was extracted and mounted on a photographic emulsion for autoradiography. The labelled DNA exposes the film producing a diffused profile of DNA molecule. The duplicating molecule showed a replication fork. After the replication, the radioactivity was found incorporated in only one strands of DNA and in the second generation, both the strands become labelled and this will be clearly seen on the photographic emulsion because of the presence of black dots.

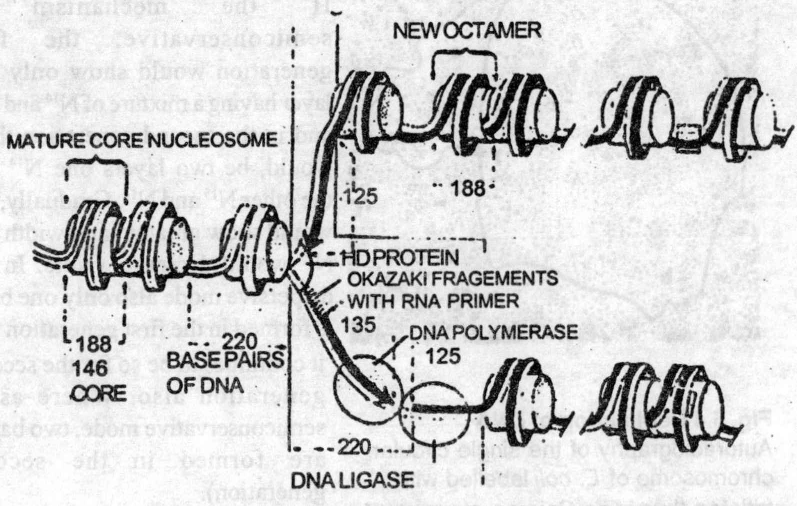

Fig. 5.5 Replication of DNA
Summary of the major steps in DNA replication

This experiment also clearly shows the mode of replication is semiconservative because after the first replication, each molecule will have only one labelledand the other non-tritiated strand. In the second generation, both the strands are tritiated in some molecules and in the other hybrid – one tritiated other non tritiated strands occur. and the other non-tritiated strand. In the second generation, both the strands are tritiated in some molecules and in the other hybrid – one tritiated other non tritiated strands occur.

Mechanism of replication

The mechanism of DNA replication in prokaryotes is as follows –

1. *Stage of replication during cell division* – It has been observed that

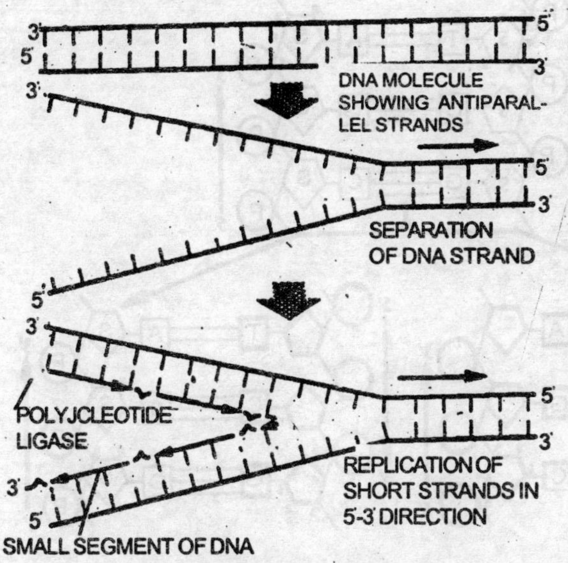

Fig. 5.6 Replication of DNA
Diagrammatic representation of discontinuous synthesis of DNA

replication takes place during interphase between two mitotic cycles. During interphase, the amount of DNA doubles due to synthesis. However, DNA synthesis does not take place during the entire interphase. It is confined to the S phase.

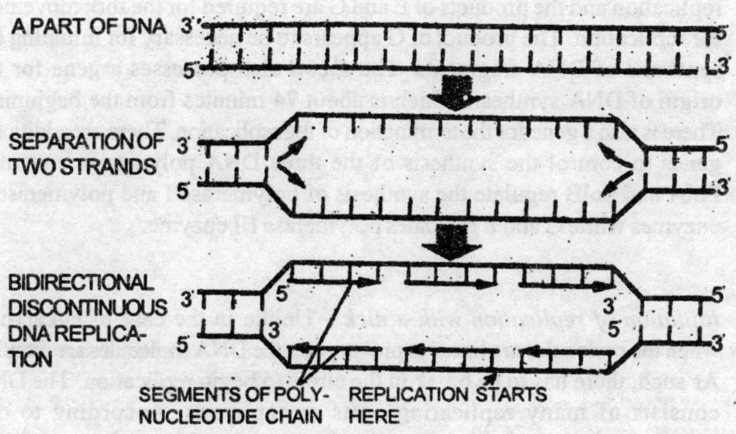

Fig. 5.7 Replication of DNA
Diagrammatic showing bidirectional DNA replication

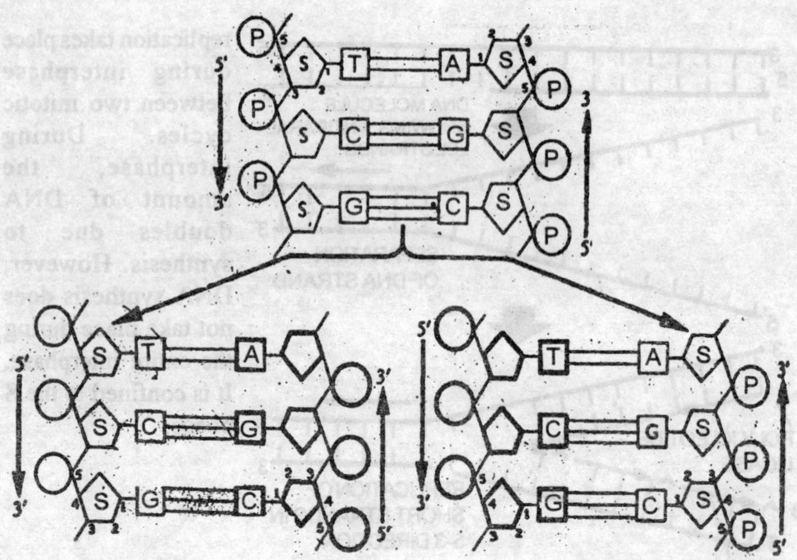

Fig. 5.8 Replication of DNA

Replication of DNA molecule. The two strands of DNA separate and each of them synthesizes its counterpart, thereby forming two similar molecules of DNA

2. *DNA Synthesising Genes* – In *E.coli* the overall process of replication requires the products of the genes *dnaA, dnaB, dnaC-D, dnaE,* and *dnaG.* The gene products of AC and D are required for the initiation of fresh replication and the products of E and G are required for the fork movement for replication. The product of G appears to be necessary for initiating the synthesis of DNA fragments. The *E.coli* also possesses a gene for the origin of DNA synthesis which is about 74 minutes from the beginning. There is also a gene for the termination of the replication. There are additional genes to control the synthesis of the three DNA polymerase enzymes. PolA and polB regulate the synthesis of polymerase I and polymerase II enzymes while G and E regulates polymerase III enzyme.

3. *Initiation of replication with a nick* – Unlike in the case of eukaryotes when the molecules are linear in prokaryotes the DNA molecules are circular. As such, there has to be break in the circle to begin replication. The DNA consists of many replicating units or replicons. According to one hypothesis, replication is initiated only after a break or nick in one of the two parental strands. The cut is presumed to take place in the helix at a

specific initiation point. This cut or incision is made by a specific endonuclease enzyme.

4. *Origin of replication* – Replication does not begin randomly, it always begins with a specific site called ori site and is regulated by a gene and its product. It has been discovered that in *E.coli*, the origin of replication point as about 74 minutes on the map.

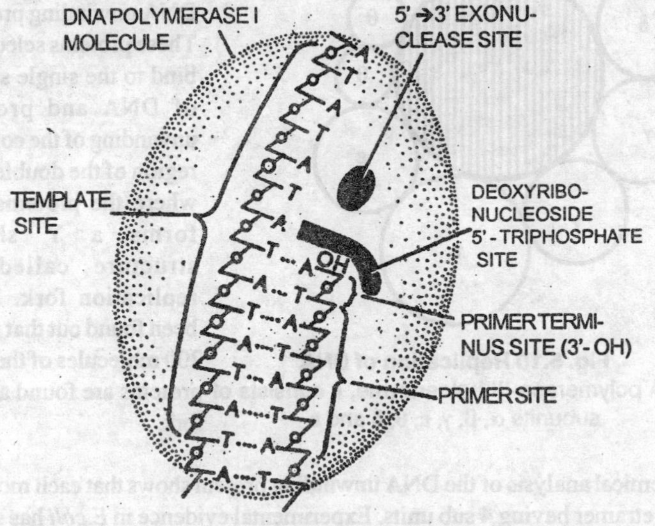

Fig. 5.9 Replication of DNA
A molecule of DNA polymerase-I showing specific binding sites

5. *Role of plasma membrane* – There are evidences to indicate that *E.coli* consists of an enzyme complex at the point of attachment of DNA to the membrane. This point of attachment is known as the replicating point. The DNA thread moves through the replicating point and duplicates itself in the process. In this process, the replication is exactly opposite of what is seen in eukaryotes where it is the enzyme that moves along the strand. The exact relationship between the plasma membrane and the DNA is not known. Quite probably, the membrane lipids may have some role to play in the process of initiation. Polymerase II and polymerase III enzymes are seen at the point of attachment.

6. *Strand unwinding* – The DNA helix has to first unwind in order to separate the two strands for beginning the replication. A study of the unwinding process in T_2 bacteriophages shows that each DNA molecule has to undergo 20,000 rotations in unwinding during replication. As the molecule itself

consists of 2 lakh nucleotide pairs, in these phages, replication must go on simultaneously along with unwinding.

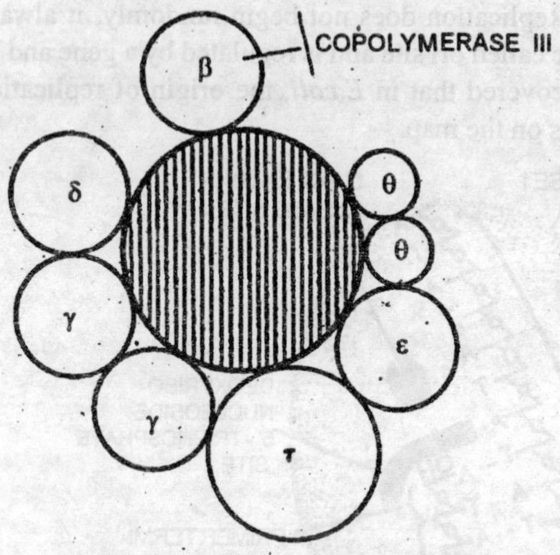

Fig. 5.10 Replication of DNA
DNA polymerase-III holoenzyme, it consists of subunits α, β, γ, ε, θ, δ and ε

The unwinding of the helix is controlled by enzyme proteins commonly referred to as DNA unwinding proteins. These proteins selectively bind to the single strands of DNA and promote unwinding of the coil. The region of the double helix where the proteins bind forms a Y shaped structure called the replication fork. It has been found out that nearly 200 molecules of the proteins are found at each fork.

Chemical analysis of the DNA unwinding protein shows that each molecule is a etramer having 4 sub units. Experimental evidence in *E.coli* has shown that the addition of DNA unwinding proteins to the medium increases the phase of the unwinding of the DNA molecule.

7. *Role of Swiveling Protein* – As the replication fork moves downwards along the parental strands, unwinding begins. For every ten nucleotides, there will be one rotation of unwinding. As the original configuration of the molecules is in the form of a wound, helix in circular DNA molecules present in prokaryotes untwisting or unwinding imposes a strain. To some extent, this strain can be received by additional twisting called super twisting in the unreplicated part of the DNA. However, this super twisting has a limit and ultimately even the supertwisting should be unwound for the replication fork to continue. The relieving of the strain imposed by unwinding in brought about by proteins called superhelix relaxing proteins. These proteins introduce small cuts in the non-replicating region of DNA thereby making one strand to rotate upon the other. Thus, relieving the strain. The breaks can then be healed by ligation.

8. *DNA Template* – To synthesise a fresh strand of DNAS apart from the

enzymes, the most important requirement are the nucleotides and a DNA template. The 4 nucleotides are adenine, guanine, cytosine and thymine. The previously existing DNA can serve as a template on which can be assembled fresh nucleotides.

The template DNA can be either single stranded or double stranded. Single stranded DNA is found naturally in some bacteriophages. Template DNA acts as a mould on which new strand of DNA is synthesised by complementary base pairing.

9. *RNA Primer* – DNA synthesis cannot be initiated on a DNA primer. DNA polymerase enzymes can initiate DNA synthesis only if there is a RNA primer. In other words, synthesis of a fresh strand can begin only if there is a small piece of a strand already. This small piece which is RNA is called RNA primer. The RNA primer is a short polynucleotide chain synthesised by the DNA template close to the point of replication. In all probability, this is catalysed by a modified RNA polymerase enzyme. It has been noticed that in *E.coli*, RNA polymerase is necessary to begin the replication.

10. *Elongation of chain* – Synthesis of new DNA strand takes place by the addition of fresh nucleotides to the 3'-OH group of the last nucleotide in the RNA primer. This synthesis takes place in 5'→3' direction and the enzyme that catalyses this is DNA polymerase III.

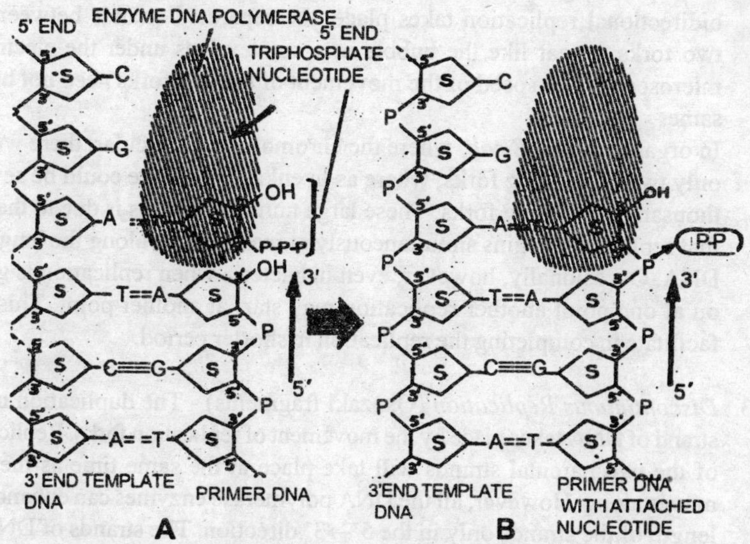

Fig. 5.11 Replication of DNA
Diagram showing action of DNA polymerase in DNA replication

The newly synthesised DNA chain has an RNA primer attached to its 5' end. After the synthesis of the chain, the RNA primer is hydrolysed by the exonuclease activity of DNA polymerase I. The resulting gap left by the RNA primer is filled by DNA nucleotides by the catalytic activity of DNA polymerase I. The freshly formed nucleotide chain is joined to the existing chain by the lygase enzyme.

Several opinions have been put forward to account for the presence of an RNA primer to initiate the synthesis of DNA chain. According to one opinion, if any errors occur during the initial synthesis, it can be eliminated when the RNA primer is removed.

11. *Direction of Replication* - It has been seen that in prokaryotes the replication may proceed in one or both directions from the point of origin. Thus, replication may be unidirectional or bidirectional. Bidirectional replication has been seen in *E.coli, Bacillus subtilis, Salmolella, Typhimirum, E.coli* phages etc. Unidirectional replication has been reported is bacteriophages P2 and 186, and also in the mit DNA of mouse LD cells. Bidirectional replication has an advantage in that, the origin is completely duplicated before the complication of replication.

12. *Formation of Replicating Forks* – At the point where the two strands get separated to initiate replication, a replicative fork is found. The shape of the replicating fork is in the form of the English alphabet Y. When bidirectional replication takes place, the separated strands between the two forks appear like the bubble or a convex lens under the electronic microscope. The speed of the movement of the two forks need not be the same.

    In organisms like *E.coli,* where the chromosome is circular, there will be only two replicating forks, where as in eukaryotes there could be several thousand replicating forks. These large number of forks is due to the fact that replication begins simultaneously at many points along the length of DNA. Occasionally, however, even in bacteria when replication is going on at one point another replication may start at another point. This will facilitate in completing the replication in shorter period.

13. *Discontinuous Replication* (Okazaki fragments) - The duplication of the strand of DNA takes place by the movement of replication forks. Replication of the two parental strands will take place at the same time as the fork moves along. However, all the DNA polymerase enzymes can enhance the length of the strands only in the 5'→3' direction. The strands of DNA on the other hand, polarise in opposite directions. This naturally poses a question as to how can DNA polymerase bring about simultaneous

replication. While the enzyme moves forward on one DNA strand, simultaneously it cannot move backward on the other.

Okazaki et al (1968), have provided an answer to the above problem. According to them only one strand 3'→5' is continuously replicated, while the other 5'→3' replicates in a discontinuous manner and synthesis, only short fragments of DNA. These short fragments have been called Okazaki fragments. The direction of synthesis of Okazaki fragments in opposite to the movement of the replicating fork. Eventually, Okazaki fragments are joined together by the enzyme polynucleotidfe ligase. In prokaryotes, the Okazaki fragments are as long as 1,000–7,000 nucleotides.

According to another suggestion (Okazaki et al), both the strands may replicate discontinuously and the fragments formed may join up later.
Evidence obtained from *in vitro* culture of *E.coli,* have shown that both the strands are synthesisd discontinuously.

Enzymes involved in DNA Replication

1. **DNA Polymerases** – Kornberg (1957), isolated an enzyme from *E.coli* which had the ability to polymerise the nucleotides dATP, dTTP, dCTP and dGTP to form a polynucleotide in the presence of a primer and DNA template. This enzyme was capable of synthesizing new DNA strands by proceeding in 5'→3' direction. Subsequent works of DeLucia and Carirns (1969), Knippers (1970), and Gefter et al (1969), showed that in prokaryotic cells there are at least three DNA polymerases – *poly I, pol II,* and *poly III.* Kornberg's DNA polymerase is equivalent to *Po! I.*

Polymerase I – This enzyme does not seem to be necessary as bacterial mutants lacking this are able to synthesize DNA. *Poly I* seems to be taking part in DNA repair. Structurally, poll is a single polypeptide chain with a molecular weight of 1,09,000. Electron microscopic studies of the enzyme have shown that the enzyme has a number of functional sites –

(i) A template site for binding template DNA.
(ii) Primer site for binding primer RNA segment.
(iii) A primer terminus site for 3' - hydroxyl terminus of primer.
(iv) A triphosphate site - locus for incoming deoxy ribonucleotide 5' group.
(v) Exonuclease site for 5'→3' exonuclease activity.

Polymerase II - This is a single polypeptide chain with a molecular weight of 90,000. *E.coli* cell has about 40 molecules of *pol III.* It helps in 5'→3' polymerization.

Polymerase III - Discovered by Kornberg and Gefter (1972), it is the most active among all the three enzymes. The enzyme is chiefly responsible for DNA chain elongation. It has a molecular weight of 5,5000. The enzyme also acts as 5'→3' and 3'→5' exonuclease..

DNA ligase– Also known as polynucleotide ligase, this enzyme joins the two fragments by catalysing the synthesis of a phosphodiester bond between a 3'-OH group at the end of one chain and 5'→3' - OH group at the other end of the other chain. The enzyme is necessary to join Okazaki fragments which are produced after discontinuous replication.

The enzyme has been isolated from *E.coli*. It is made up of a single polypeptide chain and has a molecular weight of 70,000. Each cell of *E.coli* is known to posses 2000–4000 molecules of this enzyme. DNA ligase has the following functions.

1. Joins Okazaki fragments during DNA replication.
2. Helps in sealing single strand nicks in DNA duplex.
3. Joins free ends of DNA (linear molecule) to form circular chromosome.
4. Joins the segments of DNA during recombination during meiosis.

MODELS FOR DNA REPLICATION (IN PROKARYOTES)

Jaboc and Brenner's Model

Studies with phase contrast microscope have shown that in the dividing cell, the sister chromosomes get oriented equidistantly inside the cell. In eukaryotic cell, this separation is achieved by the mitotic spindle. In prokaryotes like bacteria such a mechanism is lacking. Jacob and Brenner have proposed a model to explain the replication and separation of daughter chromosomes. According to this model, the chromosome gets attached to the mesosome at the point of origin of replication. After the completion of replication, a new membrane attachment site is formed adjacent to the old one. The strand which has acted as a template remains a circle, while the newly formed one is broken. One of the free ends of the newly formed chromosome attaches to the new binding site. During replication, the replicating fork and the enzyme system remains stationary while the DNA molecule moves along and gets replicated. Simultaneously a new membrane is synthesized between the two and the two genomes get separated. The newly synthesized DNA becomes circular eventually.

Carin's Model

At a specific point (origin), the DNA breaks and replication begins and proceeds in two opposite directions. As the replication proceeds, the DNA strand unwinds. This unilateral unwinding (the other cannot unwind freely) creates a pressure and it is transfered to the unreplicated part resulting in supertwisting. This results in the stoppage of replication. To begin replication once again, a 'neck' is made in one of the strands by a swivelling protein. This facilitates rotation of the strands and the break is closed subsequently.

Yoshikawa Model

This is a modification of the Cairn's model. According to Yoshikawa, the newly formed DNA strands covalently join to the ends of original chromosome.

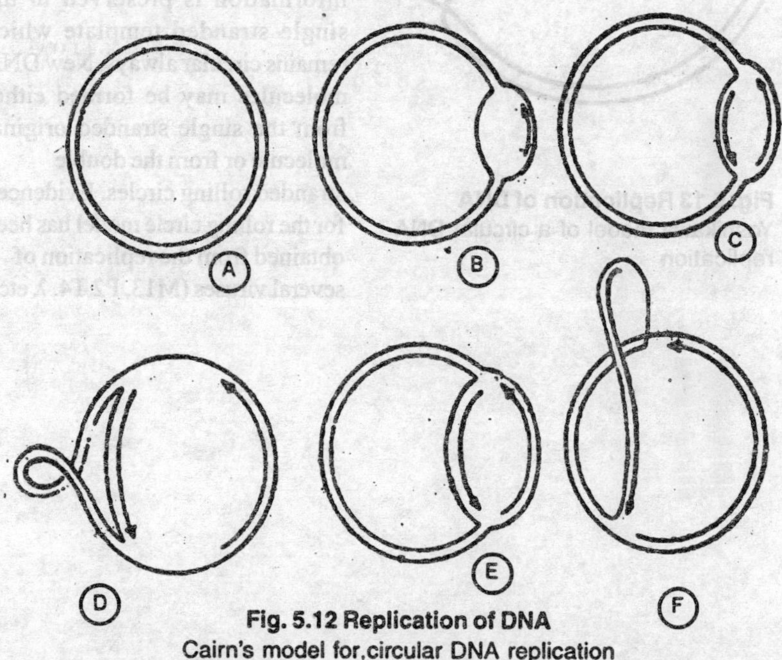

Fig. 5.12 Replication of DNA
Cairn's model for circular DNA replication

Rolling Circle Model

This is the model accepted for the replication of single stranded DNA in viruses (ϕ x174). The single stranded circular DNA becomes double stranded by the synthesis of another strand starting at the origin. One strand of this duplex ring is now cut at a point by an endonuclease. As a result of this, a linear strand with 3' and 5' end is created. The 3' end serves as a primer for the synthesis of a new DNA strand. The unbroken strand is used for this purpose; with the parental strand itself acting as a template. The 5' end of the strand becomes attached to the membrane. Meanwhile the original, unbroken single strand unwinds and

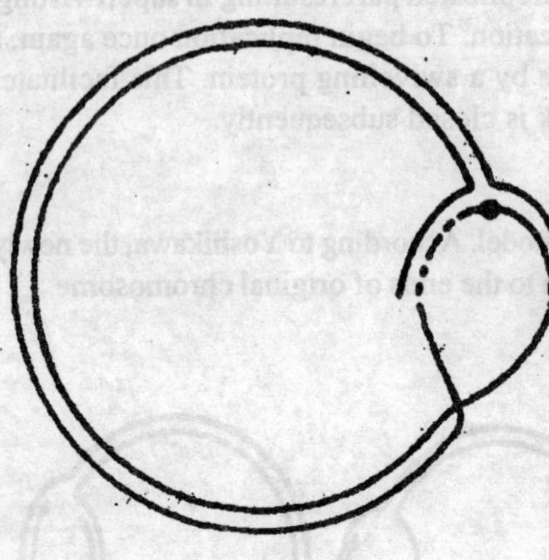

Fig. 5.13 Replication of DNA
Yoshikawa model of a circular DNA replication

separates leaving the tail attached to the membrane. The tail is cut off into a specific length – *Progeny rod.* The rod rolls and forms a circular molecule, with the enzyme ligase binding the two ends. The newly formed circular molecule which is double stranded can in turn become rolling circles. The genetic information is preserved in the single stranded template which remains circular always. New DNA molecules may be formed either from the single stranded original molecule or from the double stranded rolling circles. Evidence for the rolling circle model has been obtained from the replication of several viruses (M13, P2 T4, λ etc).

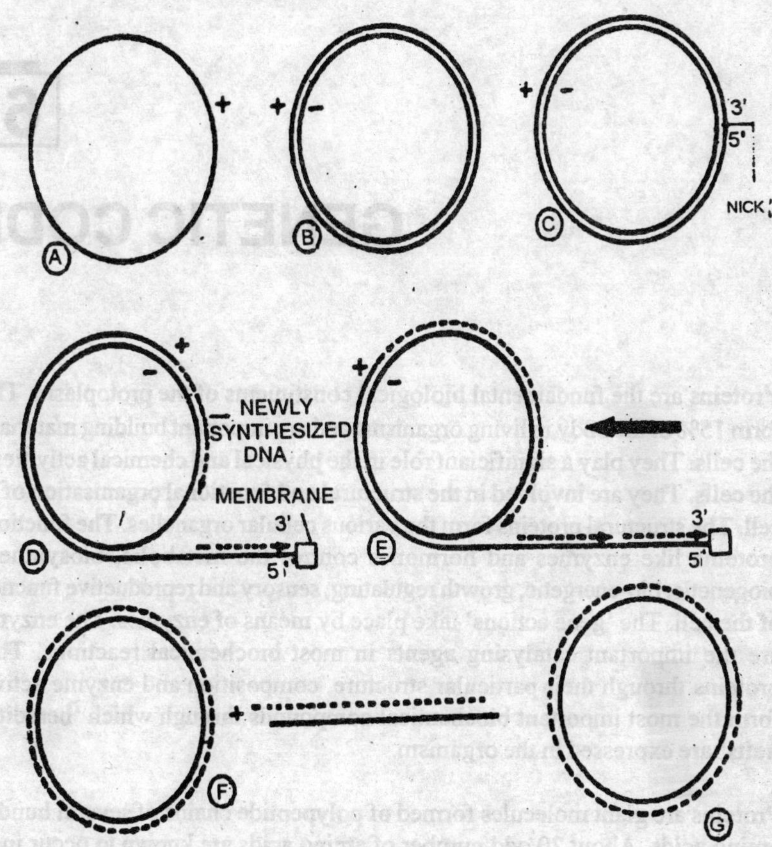

Fig. 5.14 Replication of DNA
Rolling circle model of a circular DNA replication

GENETIC CODE

Proteins are the fundamental biological constituents of the protoplasm. They form 15% of the body in living organisms and are important building material of the cells. They play a significiant role in the physical and chemical activities of the cells. They are involved in the structural and functional organisation of the cell. The structural proteins form the various cellular organelles. The functional proteins like enzymes and hormones control the metabolic, biosynthetic, biogenetic, bio-energetic, growth regulating, sensory and reproductive functions of the cell. The 'gene actions' take place by means of enzymes. The enzymes are the important catalysing agents in most biochemical reactions. Thus, proteins through their particular structure, composition and enzyme activity form the most important biochemical compounds through which 'hereditary traits' are expressed in the organism.

Proteins are giant molecules formed of polypeptide chains of several hundred amino acids. About 20 odd number of amino acids are known to occur in the cells. Amino acids are formed of a basic amino (NH_2), and an acidic carboxyl (COOH) groups. The variety of cells in an organism represent myriads of proteins. Different types of proteins are synthesised by different cells of the body. Different combinations of amino acids produce different proteins.

Synthesis of Protein

The process of protein synthesis is the most complicated biological activity. Basically, it is governed by the *genetic information* carried in the genes (DNA) of the chromosomes. Specific amino acids are linked in particular sequence or order to form a polypeptide chain of protein. The various cell organelles and chemical compounds involved in the synthetic proces are –

1. Deoxyribonucleic acid .... DNA
2. Three types of ribonucleic acids .... RNA
3. Amino acids

4. Enzymes and
5. Ribosomes

The discussion above will clearly reveal that proteins are the representative of the master molecule of the cell DNA to carry out and regulate all the metabolic activities. Proteins are important in another way also. They are important for the existence of DNA itself. A number of enzymes (proteins) are necessary for the replication of DNA itself. When proteins are of such fundamental biological importance, no wonder then the cell machinery to synthesize them (proteins) is not only complicated but precisely regulated.
Which molecule(s) present in the cell can regulate the synthesis of proteins? Obviously there cannot be another compound than DNA. The structure and properties of the DNA molecule are eminently suited to exercise such a regulation over the synthesis of proteins.

Having conceded that DNA directly is involved in the regulation of protein synthesis, we may now try to answer the two following questions – which aspect of protein structure is controlled by DNA? And how and by what mechanism DNA regulates the assembly of amino acids to synthesize the proteins?

Proteins are extremely complicated biological macromolecules. There are different dimensions for its structural complexity and integrity. These are theprimary, secondary, teritary and quaternary structures (for details – see the chapter on protein synthesis). DNA molecule specifies only the primary structure, while the other dimensions of protein structure like folding etc. is controlled by proeins themselves.

The primary structure of proteins involves the following aspects – quality, quantity and sequence of amino acids and it is this that is regulated by DNA.

What is the mechanism adopted by DNA to regulate the synthesis of proteins? Obviously, proteins cannot be synthesized *in situie* within the nucleus, because all the components needed for proteins–amino acids, ribosomes etc., are not present in the nucleus. It is not possible also for the DNA itself to migrate to cytoplasm to synthesize the proteins. The only alternative is for DNA to send a messenger or representative which carries the message to specify the quality, quantity and the sequence of amino acids to be asembled to produce the protein. This message is called the genetic code, because it specifies in genetic language the amino acids to be assembled.

The language of genetic code

A code is said to be a cryptic or hidden language, carrying a message and upon deciphering or translation, the message is understood. In the genetic code, the message is translated as the type of aminoacid to be assembled. What is the composition of the genetic code that comes out of DNA? To answer this, we must first understand the manner in which genetic information exists in the DNA molecule.

It is well known that eventhough the DNA molecule has three components namely sugar, phosphate and nitrogen bases, it is the last one that is unique to each DNA molecule in the sequence of their arrangement. The nitrogen bases vary from one DNA segment to another in arrangement, so the information is based upon the sequence of their arrangement. The sequence of nucleotides is then the code in which the DNA sends the message. How is the message sent? The DNA molecule replicates in a particular segment and the replicated segment in the form of RNA now called mRNA carries the genetic message in the form of nucleotide sequence (for details see protein synthesis). The sequence of nitrogen bases have been found to be identical to the linear sequence of amino acids in a peptide chain. In otherwords, there is colinearity between mRNA and polypeptide chain (see later in the chapter for details of colinearity).

It is thus clear that the genetic code is in the language of nitrogen bases and the reading of the code is in their sequence of arrangement.

Amino acids involved in protein synthesis

A large number of amino acids – almost upto 150 are found in nature. Of these, however, quite a number of them remain as free amino acids in the sense they do not join together to form a molecule of protein. It has been found out that only about a little more than 20 amino acids are protein amino acids in that all the known proteins are made up of only these amino acids. Some amino acids, however are found in almost all protein chains while a few others are found only in very few proteins. For example, cystine is present only in proteins which have a disulphide linkage. Hydroxylproline is another amino acid found only in few proteins.

The code is triplet (basis of crypto analysis)

A DNA molecule from every source contains only 4 types of nucleotides– adenine, guanine, cytosine and thymine. On the other hand the basic building blocks of proteins namely amino acids are more than 20 types. A basic problem that one has to analyse before finding about a genetic code is how many bases specify a unit of amino acis. Sevral opinions have been put forward in this regard.

Single code. According to one of the opinions, each nucleotide will code for one amino acid. But this would be insufficient to code for all the 20 amino acids as there are only 4 nucleotides. With 4 nucleotides only 4 amino acids can be coded. What about the rest?

Double code. This asumes that each code is formed of two letters i.e., 2 nucleotides. In other words, 2 nucleotides will specify one amino acid. In this type of a code, all the 20 amino acids cannot be coded but 16 codes (4 x 4) are possible. But there will be a number of ambiguous determination of amino acids.

Possible singlet, doublet and triplet codes of mRNA

Single Code (4 words)	Double (16 words)	Code Triple Code (64 words)
		AAA AAG AAC AAU
		AGA AGG AGC AGU
		ACA ACG ACC ACU
		AUA AUG AUC AUU
		GAA GAG GAC GAU
		GGA GGG GGC GGU
		GCA GCG GCC GCU
A	AA AG AC AU	GUA GUG GUC GUU
G	GA GG GC GU	CAA CAG CAC CAU
C	CA CG CC CU	CGA CGG CGC CGU
U	UA UG UC UU	CCA CCG CCC CCU
		CUA CUG CUC CUU
		UAA UAG UAC UAU
		UGA UGG UGC UGU
		UCA UCG UCC UCC
		UUA UUG UUC UUC

Triple code. This was first suggested by the physicist, George Gamow (1954). According to this code, 3 bases (nucleotides) will code for one amino acid. In other words, each code will have 3 nucleotides which will specify the particular amino acid. These triplets of codes are given the name codons. With triplet codes, thepossibilities would be 64 (4x4x4) codons. This would be more than the number of amino acids (which are only 20). Obviously then each amino acid will have more than one code. Experimental evidence shows that the code is a triplet and about 61 of the 64 codes are used to specify amino acids during protein synthesis.

The maximum possible number of codons in the singlet, doublet and triplet codes

Type of code	Number of bases in codon	Number of codons	Ambiguous/ degenerate
Single code	1	4	Amgiguous
Double code	2	4 x 4 = 16	Amgiguous
Triplet code	3	4 x 4 x 4 = 65	Degenerate (more than required number)

Properties of code

1. The code is non-overlapping

The DNA molecule is a very long chain and has series of nucleotides. There is a gap between the nucleotides. Naturally a question arises that if three nucleotides constitute a codon will there be any gap between two successive codes or do the codes overlap. In otherwords, is the code overlapping or non overlapping? If the code is over lapping from where will the reading begin. Because the beginning of the code reading will alter the nucleotide combination and hence will alter amino acid which is specified by the code. When Gamow proposed the code he assumed it to be overlapping. Many people attempted to rationalise the excess number of codes i.e., 64 for 20 amino acids by assuming that the codes are overlapping and the codons can be reduced to about 20. But this involves a lot of problems. If we take for instance, the codes for lysine (AGG and AAA), and codes for phenyl alanine (UUU and UUC). The code letters of these two amino acids are mutual exclusive i.e., lysine forbidden combination of amino acids and proteins. But, on the other hand these amino acids are found in proteins proving that the genetic code is non-overlapping. Further, the overlapping code assumes that a nucleotide is read twice in different combinations to code for amino acids. If this is true, the entire precision of code would be disturbed. Further, any mutational change in an overlapping code would affect more than one code thus disturbing a number of amino acids.

Mutational studies have shown that the code is non-overlapping. In TMV (Tobacco Mosaic Virus), the change of one base in the DNA due to mutation would alter the composition of protein by a single amino acid proving that the code is non overlapping. If the code is overlapping change of one base should have disturbed more than one amino acid. Studies on normal and sickle cell haemoglobins have shown that single base mutations result in the replacement of only one amino acid.

2. The code is degenerate

(Synonym codons). As it has been mentioned above, there are 64 possible codons considering the code to be a triple one. Of these 64 codes about 61 code for amino acids. As they are only amino acids for exceed the number of amino acids. This leads to the conclusion that each amino acid must have more than one code. Of the 20 amino acids, only tryptophan and methionine have a single codon each, while all the others have more than one codons. *Phenylalanine, tryosine, histidine, glutamine, asparagine, lysine, aspartic acid* and *glutamic acid* have two codons each (see the codon table). *Leucine, argenine* and *serine* have 6 codons each. Thus, there is no specificity as to the number of codons that decide an amino acid. The number of codons (whether one or many) specifying the amino acid possibly have been evolved over a long period of time during the process of evolution. The variability in the number of codons perhaps may explain the differential distribution of amino acids in proteins (those with multiple codons occur more frequently).

When there are multiple codons for an amino acid, it must be noted that these codons are not totally different. Synonym codons specifying an amino acid have the first two bases of the triplet being constant. Whereas only the third will vary. For instance, proline is coded by the following codons – CCU, CCC, CCA and CCG. It is obvious here in these synonym codons, the first two bases are common.

The multiplicity or degeneracy of the code may pose a question as to why nature has provided alternate codons for an amino acid where a single codon would have been more precise and specific. One possible reason for the degeneracy of the codes is that multiple codons protect the organism from mutations which might eliminate an amino acid altogether if there were to be only one code.

Number of codons coding for different amino acids

	Amino acids	Number of codons
1.	Tryptophan, methoionine	1
2.	Phenylalanine, tyrosine, histidine, glutamine, asparagine Lysine, aspartic acid, glutamic acid, cysteine	2
3.	Isoleucine	3
4.	Valine, proline, threonine, alanine, glycine	4
5.	Leucine, agrinine, serine	5

3. Polarity of the code

It is very essential that a particular gene should specify an amino acid always and in all situations unless it undergoes mutational changes. It is necessary therefore, that the code must be read between a fixed beginning point and a fixed end point. These points are referred to as the initiation and termination codons respectively. Another necessity is the reading of the direction of the code because any change in the direction would alter the amino acid. All these features clearly point out that the code must have a polarity i.e., a fixed start, a fixed end point and a predetermined direction. For instance, if the codon direction as given below varies, the composition of the amino acids would completely vary.

	Leucine	Isoleucine	Valine	Serine →
	UUG	AUG	GUG	UCG
←	Valine	Leucine	Leucine	Alanine

It is obvious from the above that the direction of reading the code would alter the sequence of the amino acids and obviously changes the composition of the proteins. Available molecular biological evidences indicate that the message in mRNA is read in 5' →3' direction. Based on this message, the polypeptide chain is synthesised from the direction of amino group to carboxyl group.

4. The code is comaless

One of the most interesting questions with reference to the non-overlapping code is, are there any gaps between the codes in the form of punctuation marks so that one code is separate from the other. In genetic terms this means that there could be bases between codes indicating where one code ends and the other begins. For instance, codes with intervening punctuation marks can be represented as follows –

UUG	X	AUG	X	GUG	X	UCG
Valine		Leucine		Leucine		Alanine

In the above, X can be regarded as a base separating the two codons. If a mutation occurs in any one of the codes, it will not affect other codes and consequently, the disturbance in the protein synthesised would be to the extent of only one amino acid. On the other hand, if the code is commaless (no punctuation mark-base between codes), mutation in one base would result in a drastic change of the genetic message. Researches of Dr.Khorana and others had clearly indicated that the code is commaless i.e., there is no delimiting point between one codon and the other. Khorana and his associates used artifically synthesised long polynucleotide chains which had repeating sequences to produce polypeptide chains. For example, the repeating sequences CUCUCU contains the codons for leucine (CUC), and serine (UCU). When this sequence was used for producing the proteins neither amino acid was incorporated with the protein unless the other one is also present. This clearly explains that the triplet code has to be commaless, and only then there would be alternate translation of the codes into leucine and serine.

5. Universality of the code

The code was first deciphered and worked out in microorganisms but subsequent researches have clearly shown that these codes are same in their amino acid specification from bacteria to man. Thus, the code is said to be universal. Marshall Caskey and Nirenbrg (1967), have shown by their experiments that diverse organisms such as *E.coli* (bacterium), *Xenopus laevis* (amphibian), and Guinea pig which is a mammal used the same code for their amino acyl tRNAs.

Gene mutations provide additional evidence for the universality of codes. These mutations result in the substitution of one amino acid for other. Studies have indicated, gene mutations bringing about amino acid substitutions effect the same amino acid in TMB, *E.coli* and also in man. For instance, a single point mutation, changes the amino acid in the coat protein of TMV, alpha chain of tryptophan synthetase in *E.coli,* and also in the haemoglobin of man.

Another remarkable feature of the code is not only that it universally occurs in all organisms, but it has also remained probably without any change ever since it evolved in some bacteria a few billion years ago. Any mutation that occurs would change the composition of mRNA, and consequently the amino acid composition in the protein. Changes such as these apparently have proved to be harmful to the organisms and as such there will be a strong selection pressure against the occurrence of such mutations. Hence, natural selection has persisted with the codes perhaps ever since they originated.

Ambiguity in the genetic code

There is no evidence to indicate that the genetic code is ambiguous *in vivo.* Ambiguity means that a single codon may code for more than one amino acid. The only exception to the non-ambiguity is the AUG codon in prokaryotes–it codes for formyl methionine at the initiation site while at other positions it specifies methionine. It is not clear as to how this distinction is always met precisely but it is believed that the secondary structure of the mRNA at the initiation site may have got something to do with the coding for formyl methionine. In *in vitro* conditions, however, genetic codes sometimes appear to be ambiguous. For example poly-U codon is known to incorporate some amounts of leucine in addition to phenylalanine. Treatment of ribosomes with streptomycin is known to introduce ambiguity to codons. For example in addition to coding for phenylalanine UUU is also known to code isoleucine, serine and leucine.

Codons and anticodons

During the process of translation, the message carried in the mRNA would be converted into the sequence of amino acids which are picked up by tRNA. This is achieved by the process of binding. For this binding to take place, obviously then there should be a specific relationship between the nucleotide. Sequence of mRNA is called the codon and that of tRNA is called anticodon. The reading of mRNS is in the 5'→ 3' direction and the codons are also written in the same direction. For example, codon AUG is written as 5' AUG 3'. The corresponding anticodon and tRNA should therefore be represented as 5' CAU 3'. In such a configuration, the first nucleotide of both codon and anticodon would be the ones at the 5' end and third nucleotides would be at the 3' end.

Very often, the anticodon is written in the 3' → 5' direction so that it will be easy to correlate between the bases of codon and anticodon.

Initiation codons

In most of the polypeptide chains, the first amino acid which begins the chain is methionine in eukaryotes and N-formlymethionine in prokaryotes. Methionine or N-formylmethionine tRNA complex specifically binds to the initiation sites in mRNA containing the codon AUG which codes for methionine. The AUG code, therefore, is referred to as the initiation codon. Less frequently, however, GUG (code for valine) may also serve as the initiation codon in bacterial protein synthesis. In the bacteriophage MS2, GUG is the initiation codon for the protein referred to as A protein. Researches have indicated that the GUG may replace AUG as initiation codon when the latter is lost due to chromosomal abbreation. It has also been found out that chain initiation by GUG is not efficient as AUG possibly because the former has less affinity for formyl methionyl tRNA.

Termination codons (nonsense codons)

It has been mentioned above that AUG (occasionally GUG) initiates the synthesis of a protein chain. The natural question would be which codon specifies or signals the end of termination of the synthesis of the protein chain. It has been found out that three of the 64 codons do not specify any tRNA and as such they cannot assemble any amino acid. As such, these codons whenever they are found in mRNA would signal the end of the chain. They are called termination codons as they terminate the synthesis of the polypeptide chain.

They are also called *nonsense codons* as they donot specify any amino acid. The nonsense codons are UAG (*amber*), UAA (*ochre*), and UGA (*opal*). Of the three termination codons, UAG was the first termination codon to be discovered. It was named amber after a graduate student in Germany, named Bernstein (means amber in German). This graduate student helped in the discovery of the class of mutations. The other two termination codons were obviously given the name of the colours only to maintain uniformity.

Termination codons found in mRNA will not specify any amino acid and as such stop the process of translation and bring about the release of the peptide chain from ribosomes. No tRNA molecules have anticodons complementary to the termination codons found in mRNA. Termination codons seem to have some affinity to proteins called release factors which bring about the release of the protein chain. In prokaryotes, there are three release factors RF-1, RF-2, and RF-3. RF-1 recognises the amber and ochre codons, while RF-2 recognises ochre and opal codons, RF-3 stimulates RF-1 and RF-2 ultimately releasing the newly synthesised peptide chain. In eukaryotes there is a single factor (RF protein) which is capable of recognising all the three termination codons.

Evidence for genetic code (deciphering the code)

Earlier to the pathbreaking experiments of Nirenberg, Matthaei and H.G. Khorama, work on the genetic code was largely speculative and restricted to development of theoritical concepts. The great physict George Gamow, who proposed the triplet code gave the comprehensive theoretical concept for various aspects of genetic code. The main features of his proposals are –

1. The code is a triplet.
2. Translation is a direct procedure between codon and amino acid.
3. Translation of codes is of overlapping type.
4. Code is degenerate.
5. Nuclei acid and the peptide chain are colinear.
6. Code is universal.

The works of Brenner (1957), showed that the triplet code is of non-overlapping type. Gamow's idea of the direct translation process became unacceptable when Crick proposed adoptor hypothesis. As per this hypothesis, during the translation process some mediator molecules intervene between amino acids and nucleic acids. It is now known that the intervening molecules are tRNA.

Experimental evidence for the existence of genetic code may be obtained by two basic approaches – *in vitro* approach and *in vivo* approach. In *in vitro* method, nucleotide sequence are compared with the amino acid residues of the proteins in a culture medium. In *in vivo* method, proteins and nucleic acids are obtained from intact cells and then compared.

The *in vitro* approach has the following four methods :

1. Genetic analysis of deletion mutatants of a cistron.
2. Incorporation of amino acids by synthetic polyribonucleotides.
3. Binding of known triplets of aminoacyl tRNA.
4. Amino aicd sequence generated by mRNA transcribed from synthetic tRNA of known nucleotides sequences.

1. Genetic analysis

Genetic analysis of deletion and insertion mutants (spontaneous as well as induced) in phage T4 was reported by Crick *et al* in 1961. This deletion or insertion of a nucleotide in a DNA chain leads to a frame shift mutation after the point of deletion or insertion. As a result of this, the reading of the codon subsequent to the mutation gets altered. Obviously then, the amino acid composition of the resultant protein would be different. As a result, the protein may be functionally inactive. All the mutations tested by Crick *et al,* produced a phenotype called mutant II. It was further observed by Crick *et al,* that a series of three mutations was necessary to supress the effects of a previous mutation. In other words, three supressor mutations would nullify the effect of three frame shift mutations and bring back the original phenotype. This is a clear evidence to show that the altered codon reading would return to the normal state after three point mutations thus indicating that the codon must have three nucleotides. The study also indicates that the code is commaless. If there were to be a punctuation mark between the two codes, a deletion or insertion of a base would change the number of bases between two commas adjacent to the point of insertion or deletion.

2. Incorporation of amino acids by synthetic polyribonucleotides

Synthetic polyribo nucleotides were used as mRNA by Nirenberg and Matthai to determine the meaning of codons. They produced a synthetic RNA by using the enzyme polynucleotide phosphorylase discovered in 1955 by Grunberg. Manago and Ochoa. This enzyme does not use DNA as a template and can incorporate ribonucleotides in a random order. The first synthetic RNA used for protein in a self-free system (the *in vitro* system consists of ribosomes, tRNA's and other necessary factors for protein synthesis) was poly U i.e., the mRNA consists of only one nucleotide namely, U in a series. This can be called a homopolymer as it is a polymer of only one nucleotide. The poly U mRNA was able to synthesise a homopolymer of a polypeptide chain containing only phenylalanine residues. This proved that the triplet code UUU specifies phenylaline. Subsequently, many such homopolymer species of mRNA were synthesized and were directed to produce polypeptide chains. For example, AAA was shown to code for lysine and CCC specified proline.

Nirenberg and Ochoa subsequently used heteropolymeres of mRNA to produce polypeptide chains. A heteropolymer includes a mixture of two, three or all the four nucleotides in mRNA. This yielded further information on the possible codons for certain amino acids. For example, poly CG directs the incorporation of arginine and alanine in addition to the proline coded by poly C. Some additional information on the assignment of codons to individual amino acids was obtained by varying the ratios of different nucleotides in the synthetic mRNA. In this experiment, the code assignment was at best ambiguous since all the four bases were not required for the incorporation of any of the amino acids.

3. Binding of known triplets to aminoacyl tRNA

A triribonucleotide cannot direct the synthesis of a polypeptide but it leads to binding of specific aminoacyl tRNA to the ribosomes. Nirenberg *et al,* used this property for assigning specific codons of specific amino acids. Their technique is known as Nirenberg – Leder technique. Conventionally, a trinucleotide with 5' terminal phosphate are more active and those lacking this phosphates in inducing the binding of aminoacyl tRNA to ribosomes. On the otherhand, those with 3' terminal phosphate are inactive. This shows that the codons must be triplet. In the Nirenberg - Leder technique, trinucleotides with 5' terminal phosphate are added to a cell free system containing a mixture of 20 amino acids bound to their respective tRNA. One of the 20 amino acids had radioactive carbon. For different experiments, different labelled amino acids were used but with the same trinucleotide after the synthesis of protein, the cell free mixture containing ribosomes were analyed to detect radioactivity confirming the binding of aminoacyl tRNA to the ribosomes. These studies showed that UUU, CCC and AAA induced the binding of phenylalnine, proline and lycine respectively to ribosomes. Using this technique Nirenberg *et al,* assigned most of the codons to their respective amino acids.

4. Amino acid sequence generated by mRNA transcribed from synthetic tRNA of known nucleotides sequences

H.G. Khorana and his associates used the above technique for codon assignment in *in vitro* conditions. Known as Khorana technique, in this method long specific DNA sequences, as first synthesized. These are then transcribed to mRNA molecules using the DNA directed RNA polymerase enzyme. These mRNA molecules which have known sequences of nucleotide are directed to synthesize polypeptides. The polypeptides were then subjected to sequential degradation to know the amino acid sequences. The nucleotides sequence is then compared with the amino acid sequences. The nucleotides sequence is to deduce specific codons for different amino acids. Using this technique, almost all the codons for amino acids were designated. In this technique, Khorana and commaless nature, non overlapping and triplet nature also.

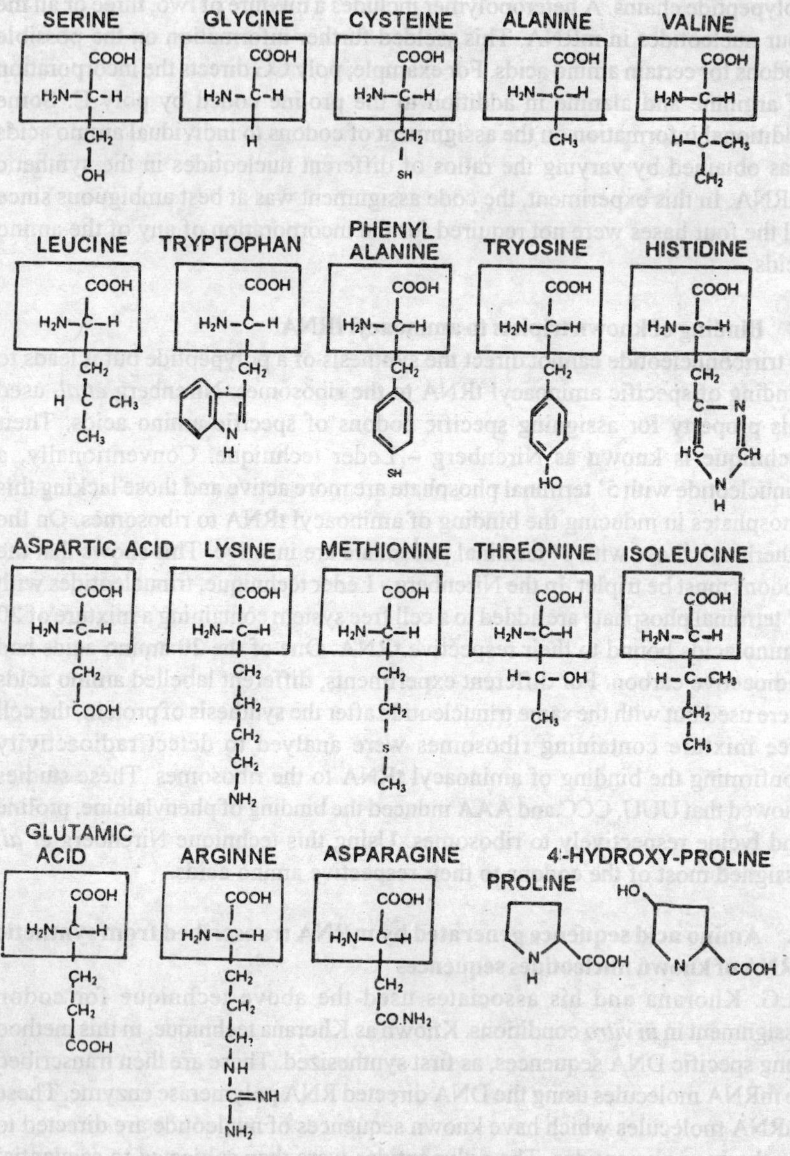

Fig. 6.1 Genetic code
The codon dictionary

In Vivo codon assignment

In the cell free systems, genetic codes have been deciphered with reference to their amino acid affinity but a doubt may arise whether these codes are the same in living systems also. In otherwords, whether the living systems use the same codes for protein synthesis. In order to prove the similarity of codes in both in *vitro* and in *vivo* three different techniques were used by three different molecular biologists.

(i) Amino acid replacement studies
(ii) Frame shift mutation studies
(iii) Comparison of a nucleotide sequence of a cistron and the amino acid sequence of the polypeptide produced by it

Amino acids replacement studies on the synthesis of tryptophan synthetase in *E.coli* (Yanofsky, 1963) have clearly indicated that the codons for the replaced amino acids, and for the amino acid replacing it prove the specificity of a codon for a particular amino acid. Frame shift mutations on lysozol enzyme of T4 bacteriophages have also shown the affinity between a specific codon and a particular amino acid.

The direct in *vivo* evidence for the genetic code comes from a comparison of the base sequence of a cistron and the amino acid sequence in a polypeptide chain produced by it. Such a study has been done with RNA bacteriophage MS2 which has only three cistrons. One of these cistrons produces the coat protein having 129 amino acids. The cistron responsible for it has 387 nucleotides. There is a total agreement between amino acid sequence expected from the base sequence of the cistron and the actual amino acid sequence of the head protein. In othewords, each codon of the cistron precisely designates the same amino acids as expected from the coding dictionary.

Wobble hypothesis - for details see protein synthesis

Colinearity of the gene structure and its product

The gene is an unbranched chain of nucleotide pairs in a particular sequence. The polypeptide chain is also unbranched and is made up of amino acid residues attached in a linear series. Thus, a linear relationship between the gene and the polypeptide chain produced by it can be indicated. Yanofsky *et al* (1964), established this colinearity in their study of tryptophan synthetase enzyme of *E coli.*

The enzyme tryptophan synthease is made up of two chains α and β. A genetic map of the α chain gene was established by determining the recombination percentage of mutations for the amino acid tryptophan. The complete amino acid residues of the α chain of the enzyme tryptophan synthetase was thus established. Subsequently, the location of amino acid substitutions in the chain for 10 mutations was determined. It was then found out that the order of mutations corresponded with the order of amino acid substitutions in the α chain of the polypeptide.

Sarabhai et al (1964), came to the same conclusion in their study of nonsense mutations in bacteriophage T4.

Pattern of the genetic code

A distinct pattern emerges when one analyses the amino acid structure and the type of codons that specify them (fig 13.1). Amino acids with similar structural properties tend to have related codons. For example, codons for aspartic acid (GAU and GAC) are similar to the codons for glutanic acid (GAA and GAG). Same relationship is found between the codons of phenylalanine, tyrosine and tryptophan – all these begin with uracil. It is believed that this feature of the code is evolved to minimise the consequences of mistakes that may arise during translation and base substitutions due to mutation. In otherwords, if an amino acidin a protein is replaced by another with almost similar properties, the protein may still remain functional.

Another pattern that is seen in the code is that for synonym codons (multiple codes for thesame amino acid), the first two bases are always constant whereas only the third one varies. The flexibility of the third nucleotide of the codon is another method to help minimise the harm posed by mutations. Crick (1966) has called this wobbling and proposed a hypothesis called Wobble hypothesis (see protein synthesis for details).

New genetic codes in mitochondria and ciliate protozoa

It has been mentioned above that the code is universal and ever since it originated it has remained static. However, during 1980s it was observed that the genetic codes of mitochondrial DNA in yeasts may be different from the others. In yeast, mitochondria UGA codes for tryptophan, while in nuclear genes in the same organism UGA is the termination codon. This, however, is not to say that in all mitochondria of all organisms, the UGA codes for tryptophan. For example in maize cytochrome, oxidase tryptophan is coded by CGG. This code in nuclear genes codes for arginine. There are also indications that AUA which codes for isoleucine in nuclear genes codes for methionine in *mit* DNA of mammals, yeast, Drosophila etc.

A different genetic code in DNA is perhaps necessiated because of the small *mit* genome size. T.H. Jukes (1966), presented a concept called *archetypal code* which will allow one anticodon (of tRNA) to pair with 4 different codons due to wobbling at the first base of anticodon and the third base of codon.

In 1986 a different genetic code were shown to be working in a ciliate member of Mycoplasma (*mycoplasma capricolum*). In the genetic code of this organism, UAA and UAG are not nonsense codons but code for glutamine.

Evolution of genetic code

Many of the features of the genetic code like degenerase, non-overlapping commaless nature etc. have led to a suggestion that the present version of the genetic code has evolved from a more primitive version which are perhaps a double than a triplet. It is believed that this shifting from doublet to triplet occurred very earily in the origin of cellular life. It is not certain, however whether all the codes for all the amino acids were evolved simultaneously or the codon dictionary gradually arrived at the present stage of 64 is not certain. One thing however, is certain eversince it came into existence in the present form, the biological necessity of it imposed such a selection pressure on it that it has remained unchanged upto the present day.

7

PROTEIN SYNTHESIS

Protein Structure And Classification

Of the two important macromolecules that mastermind all the activities of the living systems, proteins are one, the nucleic acids being the other. Proteins are present all over the cell and are important both structurally and functionally.

The name protein was first suggested by Berzelius (1838). *Protein* in Greek means 'First rank'. Proteins are highly complex organic molecules with molecular weight ranging upto many millions. Chemically, proteins are madeup of C, H, O and N. In addition to these major elements, some proteins may also haave sulphur, phosphorous, iron zinc, iodine etc. as components. The ratio of major elements in proteins are C= 50-55%, O=20.23%, N=15-18% and S=0-4%.

Proteins are made up of building blocks called amino acids (i.e. on hydrolysis, proteins yield these units). The idea that proteins are build up of amino acids was first proposed by Emil Fischer.

Amino acids are carboxylic acids which have atleast one COOH group and one NH_2 group. Other functional groups may also be present. There are atleast 22 different types of amino acids encountered in plants.

The amino acids are bound to one another in a protein chain by means of linkages or bonds called peptide linkages. Peptide linkages are formed between the COOH group of one amino acid and the NH_2 group of the other, with the elimination of a molecule of water. Several amino acids combine together with the peptide bond and form a peptide chain. Several of these peptides constitute a polypeptide chain. The polypeptide chain folds and refolds in a characteristic fashion to form a protein molecule.

Peptides

These are made up of amino acids linked with CONH (peptide) bonds. Peptides are intermediate compounds produced during the synthesis or degradation of proteins.

Amino acids ——— peptides ——— proteins

Proteins

These are the final productsin the N2 metabolism, and are made up of folded peptide chains which have many additional linkages (depends upon the point of folding), in addition to the CONH bond. Proteins have a three dimensional configuration, and have a very high molecular weight reaching upto 10^4 - 10. Although proteins contain not more than 20 amino acids, they are of immense

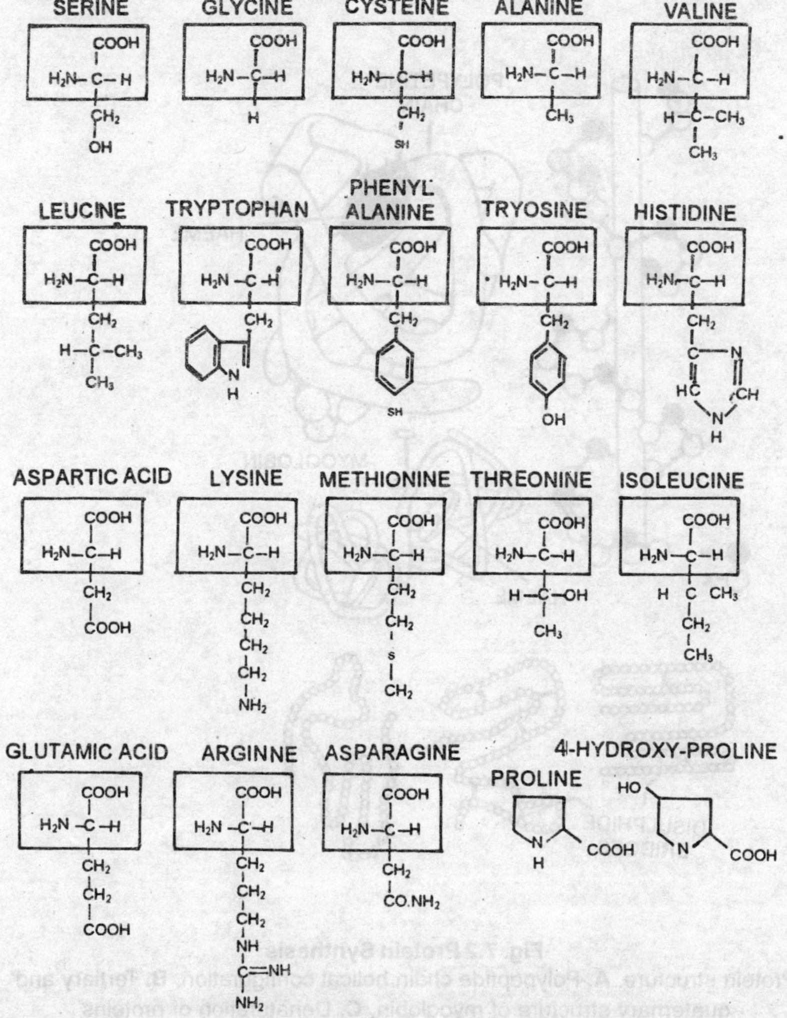

Fig. 7.1 Protein Synthesis
Structure of the amino acids

variety and complexity due to variations in combination of amino acids. In general, the three following aspects pertaining to the amino acids decide the type of protein. These are – the quality, quantity and sequence of amino acids in the peptide chains.

In other words, the type of amino acid, its quantity and the sequence in which these acids are arranged forms the basis for the categorization of proteins. Besides, proteins may differ in the number of peptide chains per molecule, type of secondary linkages (SH) etc.

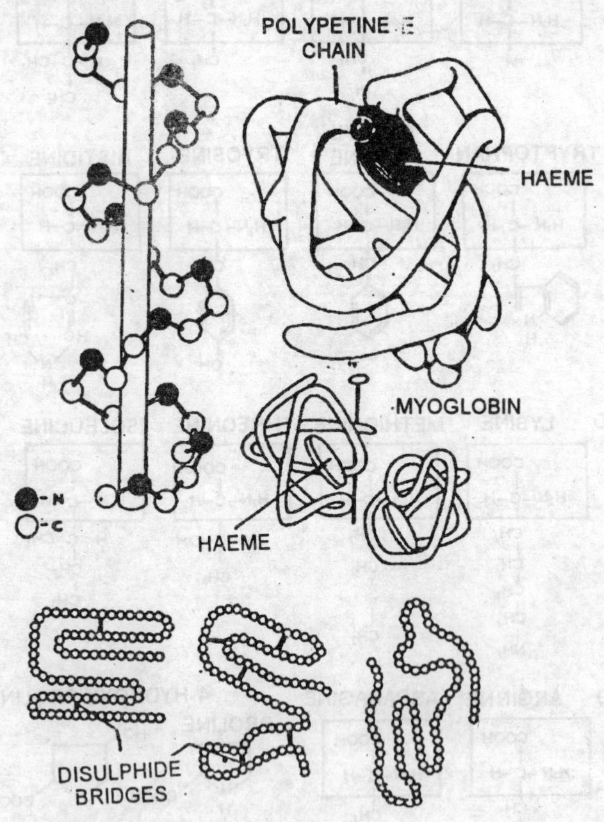

Fig. 7.2 Protein Synthesis

Protein structure. **A.** Polypeptide chain helical configuration, **B.** Tertiary and quaternary structure of myoglobin, **C.** Denaturation of proteins

Structure of proteins

Modern biochemists regard four structural levels in the protein molecule which accounts for its complexity. These levels are – primary, secondary, tertiary and quaternary structure, in the order of increasing complexity.

Primary structure

This is the simplest and the most basic structure and refers to the linear peptide chain. Besides the CONH bond, disulphide bonds (-S-S-) are found among the peptide chain linkage and the sulphur containing amino acids.

Secondary structure

The primary polypeptide chain spirally coils forming the L helical chain. The characteristic bonds of the secondary structures are hydrogen bonds between the carbonyl and amino groups of peptide links, salt linkages and Vander Wall's forces which help maintain the helical structure.

Tertiary structure

Folding and superfolding of the helical chain in a three dimensional orientation constitutes the tertiary structure. Interaction between the amino acid residues situated far apart in the chain brings about the folding.

Quaternary structure

This consists of more than one identical sub units (poly peptide chain) which associate with one another to form a functional unit. Individually, each subunit is non-functional. Such complicated structures are usually seen in enzymes. For example, the enzyme phosphorylase has two sub units, which are inactive individually, but as dimers, they are active. A protein with identical sub units in the quaternary structure is known to be homogenous or if the sub units are dissimilar, it is known as heterogenous. A protein with more than one sub unit (monomer) is referred to as an oligomeric protein.

Classification of Proteins

As the proteins possess, a very complex strucuture, their classification also poses a problem as many categorizations are possible based on structure, function etc. The following is a generalized classification. The major groups are recognized in proteins based on their solubility and chemical properties.

I. Simple proteins

II. Conjugated proteins and

III. Derived proteins.

I. SIMPLE PROTEINS

Chemically, these have only amino acids. No additional compounds are present in the molecule. Simple proteins are of the folowing kinds –

1. Albumins

These are soluble in water, dilute salts, acids and alkalies. Heat denatures the protein structure by coagulation. Some of the commonly occuring albumins are Legumelin (from legume seeds), Amylase (from barely), Leucosin (from cereals) etc.

2. Globulins

Less soluble in water than albumins but soluble in less concentrated salt solutions. Heat coagulation is seen in globulins also. Eg. Legumin (pears), Tuberin (potato) etc.

3. Glutelins

Insoluble in water and salt solutions, but soluble in weak acids and alkalies. Glutelins are generally found in plants and are heat immune. Seed proteins found in cereals like Wheat (glutenin) and rice (oryzenin) are best examples of glutelins.

4. Prolamines

These proteins are water insoluble, but dissolve in 70–80% alcohol. Prolamines have a high concentration of the amino acid proline and other amides. E.g. Gliadin (Wheat), Hordein (barley) and Zein (maize).

5. Scleroproteins

These are fibrous proteins found in integral part of the structural framework of plants. These proteins are insoluble in most of the solvents.

6. Histones

Most of the nucleoproteins, haemoglobin and globins are histones. They are insoluble in water but soluble in dilute ammonia. Histones have a high concentration of lysine and arginine.

7. Protamines

These are soluble in water, dilute acids and ammonia and do not undergo heat coagulation. Basic amino acids are rich in protamines, while tryptophan and tyrosine are completely absent. Eg. Nucleoproteins of sperms in fish.

II. CONJUGATED PROTEINS

In these proteins, non-amino acid components are associated with the protein molecule. The non-amino components are called prosthetic groups. There are seven major categories of conjugated proteins.

1. Nucleoproteins

Proteins are found in combination with nucleic acids. Hydrolysis can separate nucleic acids and simple proteins. Nucleoproteins are water soluble.

2. Glycoproteins

Also called mucoids, these are conjugated with carbohydrates like macromolecular polysaccharides, sugars, sugar phosphate etc. Glycoproteins are found in cell membranes.

3. Chromoproteins

These constitute the pigments and are coloured, the coloured appearance is due to the presence of metals such as Cu, Fe, Mg etc. Chlorophylls, flavoproteins, carotenoids, cytochromes etc. are typical examples of chromoproteins.

4. Lipoproteins

Proteins in conjugation with lipids are called lipoproteins. Lecithin, Cephalin etc. are the lipids found in combination with proteins. Lipoproteins are an integral part of all cell membranes.

5. Metalloproteins

Certain types of enzymes which require metals as activators belong to this group. Many of the respiratory enzymes belong to this group.

6. Phosphoproteins

These are proteins with phosphoric acid as the prosthetic group. Phosphoproteins are insoluble in water, but soluble in alkalies. E.g. milk and egg protein (casein and viteline).

7. Leeithoprotein

Proteins with lecithin as the prosthetic group (E.g. white of egg) belong to this group.

III. DERIVED PROTEINS

These are degradation products of proteins. Primary proteins on hydrolysis with acids or alkalies yield derived proteins.

(A) Primary derived proteins and

(B) Secondary derived proteins

A. Primary derived Proteins

There are two categories of primary derived proteins. These are Metaproteins and Coagulated proteins.

(i) Metaproteins : These are insoluble in water but dissolve in dilute salt

solutions, acids and alkalies. Hydrolysis of natural proteins on long treatment with alkalies or acids yield metaproteins.

(ii) Coagulated proteins: Treatment with heat or alcohol of natural proteins yields these proteins.

B. Secondary derived proteins

Three types of secondary derived proteins have been identified. These are proteoses, peptones and peptides.

(i) Proteoses : Prolonged hydrolysis of metaproteins yield proteoses (eg. Albuminoses from albumin). These are water soluble but immune to heat treatment.

(ii) Peptones : Peptones are obtained by continued hydrolysis of proteoses. Like proteoses, these are also water soluble and heat insensitive.

(iii) Peptides : Prolonged hydrolysis of natural proteins yields peptides. These are also water soluble and not coagulated by heat.

Functions of proteins

Most of the vital activities of the cell are controlled by the two master molecules, namely proteins and nucleic acids. Proteins are important both from the point of view of structure and function of the cell. Structurally, proteins constitute in integral part of membranes, pigments etc. Functionally, in the form of enzymes, proteins play such a vital role in cell physiology, that without proteins (enzymes), no metabolic reaction is possible. Even the synthesis of DNA is regulated by proteins. In the form of Chromoproteins (Chlorophyll, Cytochromes etc), proteins mediate all energy transformation reaction. It is because proteins are so important that the nucleic acid directly regulates the protein synthesis.

Protein structure is highly complex (as has been pointed out earlier), as well as specific. The primary structure is determined by the quality, quantity and the sequence of amino acids. Even an alteration of the position of one amino acid would alter the structure of the protein, and if it happens to be an enzyme regulating an important reaction, the entire reaction is disturbed leading to several abnormalities. Hence it is but natural that there is direct control of DNA in the synthesis of proteins. DNA, RNA and ribosomes of the cell are involved in the synthesis of proteins.

The central dogma : The basic mechanism of protein synthesis is that the nuclear DNA regulates the assembly of amino acids in the ribosomes to produce a polypeptide chain. Since DNA cannot move out of the nucleus it sends a messager – mRNA which carries the message in the form of sequence of nucleotides. This mRNA forms a template on the ribosomes. Another type of RNA, tRNA found in cytoplasm (whose nucleotide sequence is complementary to that of mRNA), picks up the amino acids and assembles them on the ribosomes to produce a peptide chain.

This may be briefly expressed as follows :
DNA Replication → DNA Transcription → RNA Translation → Proteins translation. Transcription is the copying of a complementary messenger RNA strand on a DNA strand. In translation, the genetic information present in mRNA (in the form of sequence of nucleotides) directs the sequence of amino acids to be assembled. We shall study these events in detail.

Transcription

Copying of a complementary strand of mRNA requires the following – a template, activated precursors, a divalent metal ion and RNA polymerase.

DNA molecules serves as a template for the production of mRNA. Of the two strands of DNA, only one strand called the 'sense' or the 'coding' strand transcribes mRNA. Single stranded DNA also can serve as a template.

The activated precursors required for mRNA synthesis are – Adenine triphosphate, Guanine triphosphate, Uracil triphosphate and Cytosine triphosphate (ATP, GTP, UTP and CTP)

Mg^{++} or Mn^{++} are the divalent metal ions required for mRNA synthesis.

The enzyme RNA polymerase is necessary for transcription. The nucleotides of mRNA are assembled together with the help of RNA polymerase. The enzyme consists of two parts - a core enzyme and a sigma factor. The sigma factor initiates transcription of mRNA on the DNA template and the core enzyme continues transcription. The DNA base sequence are thus transcribed into mRNA base sequence.

Genetic Code

This is the message in mRNA in the form of nucleotide sequence. Three nucleotides code for one amino acid. It is called a triplet code or codon. There are 64 codons for the 20 odd amino acids. i.e., there is more than one code for each amino acid. These alternative codes are helpful in the cae of random mutations which may alter a code (nucleotide).

The nucleotides of mRNA are complementary to the nucleotides of tRNA (present in cytoplasm). The triplet nucleotides of tRNA are called anticodons. Each tRNA has three nucleotides. The pairing between the codon and anticodon (mRNA and tRNA), follows the A-U, G-C combination. For example, if mRNA consists of the first code UUU, it pairs with a tRNA having an anticodon. AAA. Thus the series of codons of mRNA determine the series of anticodons of different molecules of tRNA, and hence the amino acids (Each tRNA molecule is specific to a particular amino acid). Since the triplets of mRNA are determined by the base sequence of DNA, it follows that the DNA molecule decides the sequence of the amino acids and hence the structure of the protein molecule.

Some triplets do not code for any amino acid (UAA and UAG). These are called nonsense codes. The so called nonsense codes are actually terminator codons and are now called ochre and amber (UAA and AUG). Code AUG is called initiator codon since it initiates the synthesis of a polypeptide chain.

The triplet codes for all amino acids collected together is called a codon dictionary. The dictionary was first proposed by Nirenberg (1965), and subsequently worked out in further detail by Khorana H.G.
The mRNA is released out of the nucleus into the cytoplasm where it is attached to the smaller sub unit of the ribosomes. Many ribosomes are connected by a single mRNA molecule, constituting a polysome or polyribosome. The mRNA attaches itself to the ribosomes in such a way, that its bases remain free (to be able to undergo pairing with the bases of tRNA). The whole structure (polyribosomes + mRNA), is called mRNA complex.

tRNA (transfer RNA) : This is present freely in cytoplasm. Also called sRNA (soluble RNA), it is the smallest among the various types of RNA molecules found in the cell. Although the tRNA is single stranded like other types of RNA, it appears double stranded because of coiling. Complementary base pairs of the same strand form pseudo bondings of hydrogen due to folding. There are two regions which are free without bonding – one at the acceptor and which always has CCA base pair sequence, and the other in the middle region of the molecule called the anticodon region, where there bases turn out of the folded portion and remain free. The types of tRNA is based on the types of these free bases. tRNA has the following functions.

(i) To attach to a particular amino acid at the acceptor, and (CCA) with adenine nucleotide. The choice of amino acid, however, depends on the type of free bases present at the anticodon end.
(ii) to attach itself at a specific site on the mRNA, and
(iii) to free the amino acid after it gets attached to another amino acid by means of peptide linkage.

The Wobble hypothesis
The triplet code or codon is degenerate i.e., there are many more codons than there are amino acids. Only 20 amino acids are involved in protein synthesis, while there are 64 possibilities or types of codons (43). Hence, more than one codon can code for an amino acid. It is interesting to find out as to why the

codes are degenerate? There could have been only so many codons as there are amino acids. But there is a problem here. If there were to be only one code for an amino acid, what happens if there is a mutation at that site in DNA? Obviously the code changes and the amino acid cannot be incorporated in the protein leading to abnormalities. On the other hand, if there are alternate codes, they act as a protection against mutation in a specific base. Hence, degeneracy of codes is a must.

Regarding tRNA present in cytoplasm, since they have to bind themselves to a specific codon, there should have been as many tRNA's as there are codons. But the number of tRNA is only as many as there are amino acids. This means, the anticodons of the tRNA's must be able to 'read' more than one codon of mRNA. How is this reading possible when bonding between base pairs is highly specific?

Crick (1966), proposed **'the Wobble hypthesis'** in order to solve this apparent dilemma. According to this hypothesis, only the first two bases of the codon have a precise pairing with the bases of the anticodon of mRNA, while the pairing between the third basis of the codn and anticodon may Wobble (non-specific). The pairing in the third base is ambigous. Thus a single tRNA can pair (bind) with more than one mRNA codon differing in only the third base. For example, the anticodon UCG of tRNA can read codons AGC and AGU of mRNA. The pairing between UGC and AGU is a perfect pairing as per the Crick and Watson pattern. But in the second case i.e., between UGC and AGU, the pairing between the first two bases is normal, while between G and U is against the normal pairing pattern. This unsual bonding G and U is called Wobble pairing.

The degeneracy of the code is not totally randon. Multiple codons for an amino acid always carry the same bases in the first two positions in a triplet with only the third 'Wobbling'. For example all the four codons for Valine have the first two bases - GU and for alanine it is GC.

Valine - GUU, GUC, GUA and GUG.

Alanine - GCU, GCC, GCA and GCG.

Translation

The process of translation consists of activation of amino acids, transfer of activated amino acid to tRNA, initiation of synthesis of polypeptide chain, chain elongation and chain termination.

Activation of amino acids

Before amino acids (20 of them) get themselves attached to the specific type of tRNA they have to be activated (energized). The activation is carried out by the ATP molecule in the presence of the specific enzyme called aminoacyl synthetase with Mg as the cofactor. The reaction produces aminoacyl adenylate (AAA) or aminoacyl AMP. Pyrophosphates are released in the process.

Amino acid + ATP + Enzyme ———— Enzyme AA> AMP complex + Ppi

Transfer of activated amino acids to tRNA : The activated amino acid (aminoacyl AMP) is transferred on to a specific tRNA, during which AMP and the enzyme (aminoacyl synthetase) are released. The amino acid gets attached to the CCA and tRNA.

Enzyme AA. AMP + tRNA ———— AA. tRNA + AMP + Enzyme

Initiation of Polypeptide chain synthesis

This involves two steps viz., preparation of the site of protein synthesis and initiation of the polypeptide chain.

(i) Preparation of the site of synthesis

Ribosomes in cytoplasm are the actual sites of protein synthesis. These are usually into their subunits (50s and 30s), when they are not participating in protein synthesis. The mRNA usually attaches to the 30s subunit through its 5' end carrying the AUG codon. In the process of attachment certain substances called initiation factors (IF) help. These factors are of three types in prokaryotes (IF1, IF2, IF3) and are of six types in eukaryotes (eIF2, eIF2al, eIF2a2, eIF2a3, eIFa4 and eIF3).

The attachment of the mRNA to the 30s submit requires the help of IF3 and Mg^{++}. The IF3 probably unwinds the rRNA (ribosomal RNA), so that it can bind to mRNA. After the attachment of mRNA, an mRNA - 30s complex is formed.

The starting amino acid in a polypeptide is methionine *(met)* in the case of eukaryotes, and N-formly methionine *(f-met)* in the case of prokaryotes. The amino acid (met or f-met) tRNA complex attaches to the initiation codon (AUG) on the mRNA 30-s complex through its anticodon (UAC). This forms the 30s initiation complex and requires the participation of IF2, IF3 and GTP (energy donor).

IF3

(i) 30s + mRNA ———— mRNA - 30s complex

(ii) mRNA - 30s complex + met tRNA ———— 70s - mRNA - met - tRNA

During the formation of the initiation complex , the 50s subunit combines with the 30s subunit to form the 70s unit. This union is facilitated by the participation of IF1. With the formation of the 70s ribosome), the initiation of the polypeptide chain begins.

(ii) Initiation of the polypeptide chain

Each ribosome has two cavities - P site (polypeptide site) and met-tRNA (aminoacyl sites). AA tRNA complex occupies the A site, while the met - tRNA (tRNA with the initiator aminoacid - methionine) will be at the P site. This occupation requires a molecule of GTP and the participation of IF_1. The energy of GTP is utilized for the formation of the peptide linkage.

Peptide linkage occurs when the carboxyl (COOH) group of peptidyl tRNA (met. tRNA) present at the P site reacts with the amino group (NH_2) of AA tRNA at the A site. The enzyme peptidyl transfers and catalyses the formation of peptide bond. Later the AA+tRNA moves from A site to the P site in the presence of GTP and IF_2

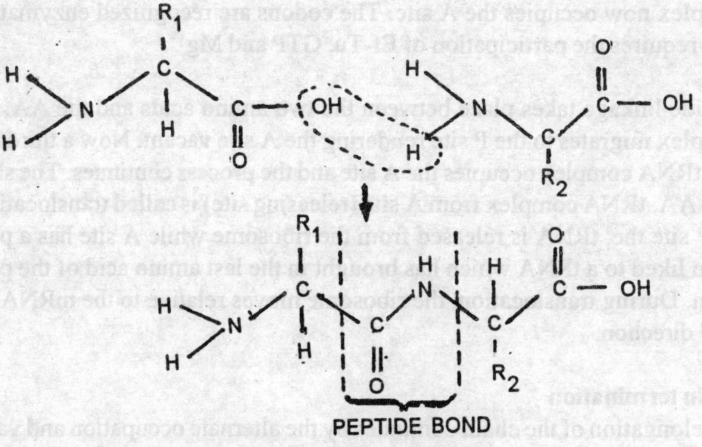

PEPTIDE BOND

Fig. 7.3 Protein Synthesis
Formation of peptide linkages

Formation of the polypeptide chain

A specific type of tRNA (which has bases complementary to the codon of mRNA), linked to a particular amino acid gets attached to a specific site on mRNA when its anticodons can form a pairing with the first codon of the mRNA. Soon after, a second tRNA with its amino acid joins the second site (triplet) on mRNA. The first and the second amino acids attached to different tRNA form a peptide linkage as both are linked to mRNA *via* the anticodons of tRNA. A molecule of water is released during the formation of the peptide linkage.

$$COOH + NH_2 \xrightarrow[\text{Peptide bond}]{} CONH + H_2O$$

As soon as the peptide linkage is formed, the tRNA is made free and comes back to cytoplasm. This whole process in which a specific tRNA carries its amino acid to a specific site is called translation i.e. the code (triplet) is translated into an amino acid. Here the sequence of bases in tRNA will be identical with the base sequence of that part of DNA coding for mRNA.

Chain elongation

Elongation of the chain requires the participation of certain factors called elongation factors (EF). These are EF-Tu, EF-TS, and EF-G in prokaryotes, and EF-1 and EF-2 in eukaryotes. After the first amino acid tRNA complex has moved from the A site of the ribosome to the P site, a second AA. tRNA complex now occupies the A site. The codons are recognized enzymatically. This requires the participation of Ef-Tu, GTP and Mg^{++}.

Peptide linkage takes place between the two amino acids and the AA. tRNA complex migrates to the P site rendering the A site vacant. Now a third amino acid tRNA complex occupies the A site and the process continues. The shifting of th AA. tRNA complex from A site (releasing site) is called translocation. At the P site the, tRNA is released from the ribosome while A site has a peptide chain liked to a tRNA which has brought in the last amino acid of the peptide chain. During translocation, the ribosome moves relative to the mRNA in the 5'-3' direction.

Chain termination

The elongation of the chain continues by the alternate occupation and vacancy at the A site by the AA. tRNA complexes until a termination codon (UAA, UAG or UGA) reaches the ribosomes. The termination codon sends signals to the ribosomes to attach release factors (RF-1, RF-2 and RF-3 in prokaryotes and RF in eukaryotes).

The release factors interact with the enzyme peptidyl transferase (which catalyses the formation of peptide bond) resulting in the hydrolysis of the (i)bond between the polypeptide chain and the tRNA (linked to the last amino acid of the peptide chain at the site) and the former is released from the ribosome.

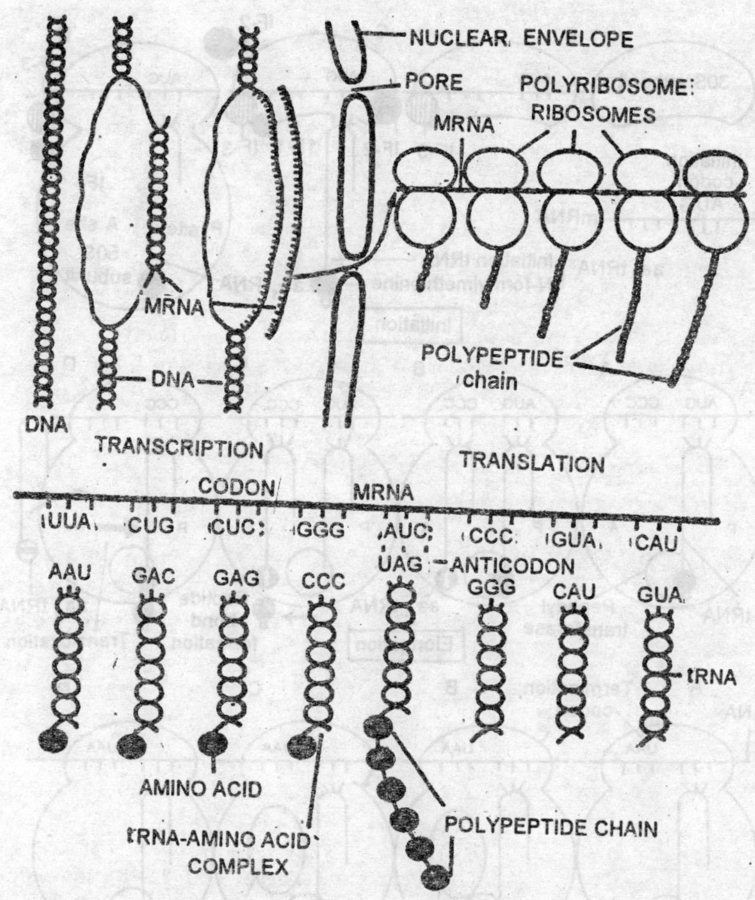

Fig. 7.4 Protein Synthesis
General scheme of protein synthesis

Other events that occur at the termination of the polypeptide chain are –

Release factors get disassociated from the ribosomes due to the hydrolysis of GTP.

(ii) tRNA is also released.

(iii) Ribosomal sub units (30s and 50s) get disunited.

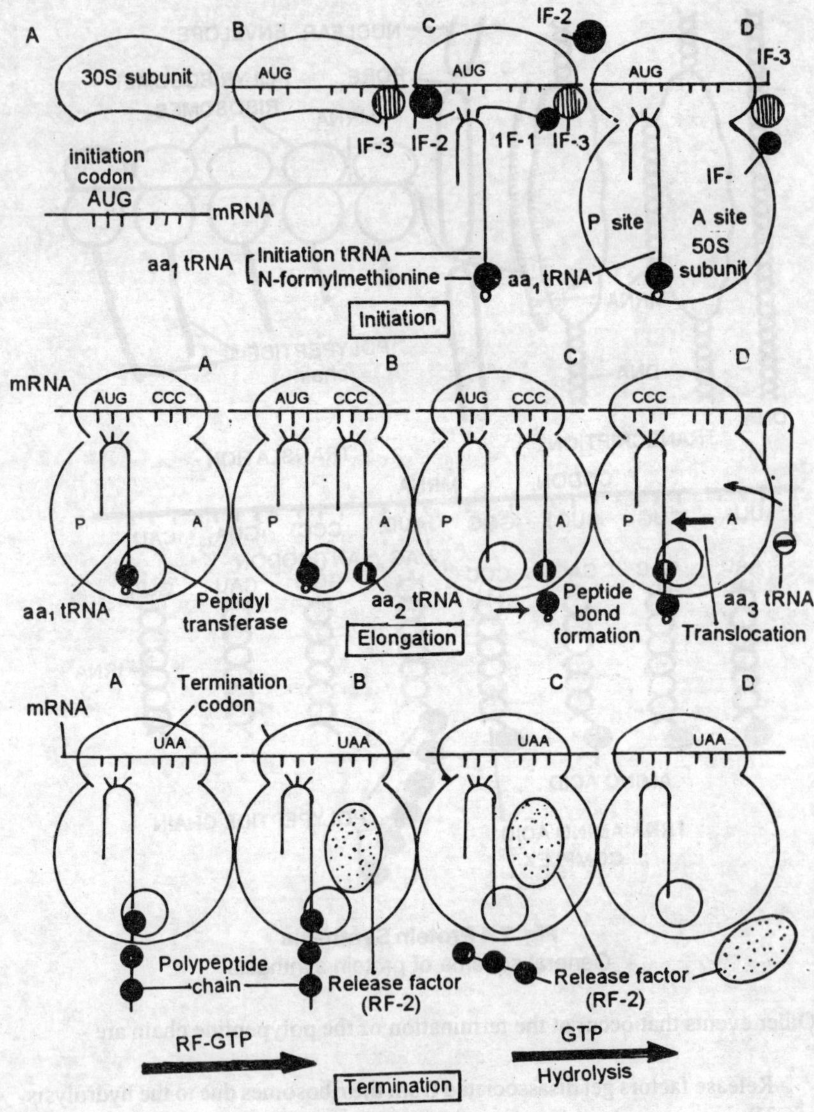

Fig. 7.5 Protein Synthesis
Different stages of translation

(iv) The polypeptide chain is processed by the removal of the methionine residue from the chain. Depending upon the circumstances, the chain may get cleaved into two or more pieces.

In eukaryotes, the polypeptide chain passes through a tunnel between the two subunits of the ribosome into the cavity of the cisterna of the endoplasmic reticulum.

Rate of Protein synthesis

The rate of transcription in *Escherichia coli* DNA for the enzyme controlling trypotophan synthesis is 28 nucleotides per second at 39°C. The speed of translation is 7 amino acids per second. For the haemoglobin molecule, the rate of translation is 4 aminoacids per second.

DIFFERENCE IN PROTEINS SYNTHESIS BETWEEN PROKARYOTES AND EUKARYOTES.

No.	Character	Prokaryotes	Eukaryotes
1.	Ribosomes involved in protein synthesis	70s, with 30s and 50s sub units.	80s with 40s and 60s sub units.
2.	Occurrence of transcription and translation	Translation begins even while transcription is transcription and still going on in mRNA	Translation begins after complete mRNA comes out of the nucleus.
3.	Migration of mRNA	Ribosomes pull out mRNA from the DNA	mRNA moves out of the nucleus through the pores in the nuclear membrane.
4.	Sites of initiation of protein synthesis in mRNA	many	one
5.	Initiator amino acid for protein synthesis.	N-formyul-methionine	Methionine
6.	Initiator factors	IF1, IF21 and IF3	eIF2, eIFwa1, eIF2a2 eIF2a3 and eIF3
7.	Elongation factors	EF-Tu, Ef-Ts and Ef-G	Ef-1, Ef-2
8.	Release factors	Rf-1, Rf-2 and RF-3	Rf (single factor)

8

GENETIC REGULATION

All the activities of an organism, be it unicellular or multicellular, are ultimately controlled by the genes. In multicellular organisms, where cells differ widely in structure and function all of them carry the same genes, since all of them are traceable to a single cell viz., zygote. This poses an interesting questioen as to the function of genes. How do similar genes function differently? The answer to active at all times. The genes are *switched* on and *switched off* at different times. This differential gene action is also responsible for the synthesis of different types of proteins (enzymes) at different times. Not all enzymes are required at all times. Enzymes required during early stages of development may not be required at later stages in the life cycle and *vice versa.* The metabolic products produced may induce or repress the enzyme synthesis.

Genetic control on protein synthesis is exercised at three stages – transcription stage, translation stage and at the release of polypeptide chain (post translation stage).

The control mechanism ensures that the cell is not unnecessarily flooded by the enzymes which are required only at particular times and only in particular quantity. Depending on the requirement the enzyme synthesis can either be induced or suppressed. Genetic regulation is seen both in prokaryotes and in eukaryotes. While the purpose of genetic regulation is same in both prokaryotes and eukaryotes, the mechanisms differ. In prokaryotes, the mechanism of genetic regulation is simple when compared with the eukaryotic organisms. The reasons for this are not far to understand. As the eukaryote organisms are far more complex they do require a very sophisticated mechanism of genetic regulation. In this chapter we will discuss genetic regulation in both prokaryotes and eukaryotes.

GENETIC REGULATION IN PROKARYOTES

Induction and Repression

In *E.coli,* the synthesis of the enzyme β *galactosidase* has been studied in considerable detail. This enzyme breaks up lactose into glucose and galactose. If lactose is not supplied to *E.coli* cells, the presence of the enzyme β *galactosidise* is hardly detectable, but as soon as the substrate (lactose) is added, the production of the enzyme β *galactosidase* increases by as much as 10,000 times. The production of enzyme again gets reduced if lactose is removed from the system. Such enzymes whose sythesis can be increased many times by the presence of the substrate are called **inducible enzymes.** The genetic system responsible for synthesis of such enzymes is called the *inducible system.* The inducible enzyme is an example of adaptive enzyme in the sense its (enzyme) synthesis gets increased due to the presence of the substrate.

In some other systems, the reverse of the inducible system occurs. For instance, in a culture of *E.coli,* if no amino acids are supplied from outside, the cells can produce all the enzymes required for the production of amino acids. On the otherhand, if a particular amino acid like histidine is added to the system, the production of histidine synthesising enzymes will fall down. In these systems the addition of a product of a synthetic pathway will check the production of enzymes responsible for the synthetic pathway. In other words increased presence of a substrate would reduce the synthesis of an enzyme which has the enhancing effect of the production of that substate. Systems such as these are called *repressible systems*, and the enzyme whose synthesis gets checked is known as the **repressible enzyme.**

Inducer and Co-repressor

It has been mentioned above that an enzyme gets induced or represented by the presence of a particular substrate. The substance which induces the synthesis of an enzyme is called the inducer and the substance which supresses the synthesis of the enzyme is called the repressor. In the synthesis of β *galactosidase,* lactose acts as the inducer and in the supression of histidine synthesising enzyme, histidine acts as the repressor.

Enzyme induction or repression is generally brought about by the substrate on which an enzyme can act. But in some instances, there will be molecules which resemble the substrate which also can act as the inducers. A typical example of this kind of induction is IPTG (isopropyl thiogalactoside). IPTG resembles lactose and can induce the production of the enzyme β *galactosidase.* Such inducers which are not substrates themselves but can bring about induction because of similarities with the substrate are known as **gratuitious inducers.**

In the above paragraphs, we have discussed as to how in the absence of lactose there is no synthesis of β *galacotosidase.* This means then in the absence of the inducer, the genes responsible for the production of the enzyme protein do not function. It will be interesting to find out as to why the enzymes do not function in the absence of lactose, and how in the presence of lactose the genes regulating the enzyme synthesis get activated. It has been found out that a group of molecules called *repressors* are present in the cell and they check the activity of the enzymes. In the presence of the inducers, the repressors get inactivated. In some instances, ordinarily the repressors are inactive but they become active in the presence of another group of molecules called co-repressors. The induction and repression is supposed to work in the following manner:

1. **Inducible system : active repressor + inducer = inactive repressor**
2. **Repressible system : inactive repressor + corepressor = active repressor**

The operon concept

F.Jacob and J.monod (1961), the Nobel laureates have studied in detail the inducible system in the synthesis of the enzyme β *galactosidase* in *E.coli.* In order to explain the induction and repression of enzyme synthesis, they have proposed the involvement of a group of genes along with the presence of cytoplasmic inducers and repressors. Their model of gene action is popularly called the operon model. This scheme of gene action proposed by Jacob and Monod as originally conceived by them consisted of an operator gene and a number of structural genes. As per the current understanding, however the operon system consists of structural gene, operator gene, promoter gene, regulator gene and repressor, inducer and co-repressor. The following is an account of the components of the operon model.

Structural genes

There are many structural genes associated in an operon system. These direct the synthesis of the mRNA and govern the sequence of amino acids in a protein molecule. Each structural gene might produce a particular kind of protein or all structural genes might regulate the production of a single protein. The activities of the structural gene(s) are controlled by the promoter and operator site of the operon system. The most well studied structural genes (z,y and *ac*), are those of the lactose operon system in *E.coli.*

Operator gene

The operator gene is situated adjacent to the first structural gene. It switches on or switches off the functioning of the structural gene (protein synthesis). In case a structural gene has to be suppressed, a repressor attaches itself to the operator to form an Operator-repressor complex. In the case of protein synthesis, the operator-repressor complex prevents the transcription by blocking the movement of RNA polymerase.

Promoter gene

The promoter gene is continuous with the operator gene and is believed to lie left to it. It is suggested that RNA polymerase binds to the promoter site during transcription. Three regions have been recognised in the promoter site. These are (i) recognition site, (ii) initial binding site and (iii) the mRNA initiation site (operator site).

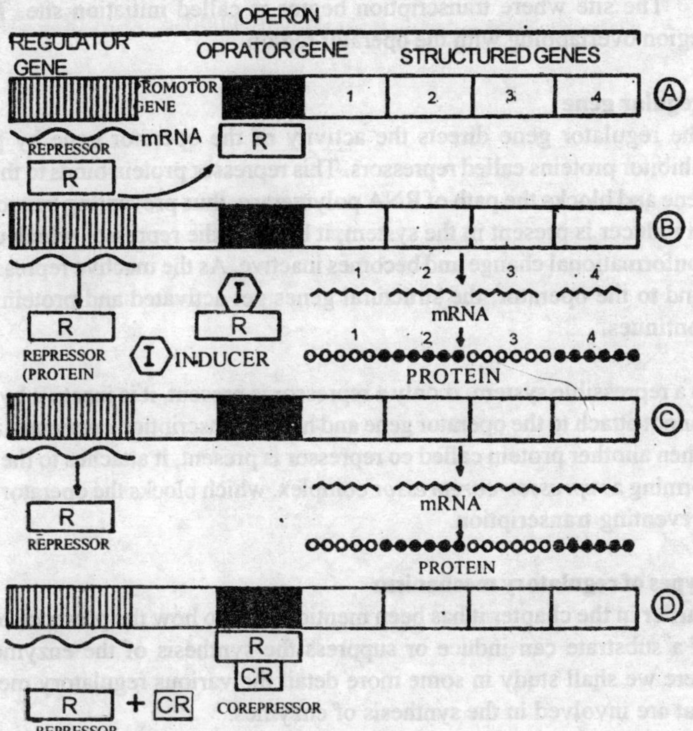

Fig. 8.1 Genetic regulation

Genetic regulation of protein synthesis in prokaryotes (the Operon system of Jacob and Monod, 1961)

(i) Recognition site

Also called the *cga* site (catabolic gene activator site), it consists of certain palindromic sequence of DNA. These symmetrical sections of DNA are recognised by proteins having symmetrically placed sub units. This site, also called the CRP site (cyclic AMP receptor protein site) binds a Carotenoids protein to the promoter gene and thus facilitiates the binding of the enzyme

RNA polymerase. It has been found that in *E.coli*, CRP combines with cAMP (cyclic adenosine monophosphate), forming a CRP + cAMP complex which binds to the promoter enhancing the binding of RNA polymerase and activates transcription. This regulation is called positive control.

(ii) Initial binding site

This consists of seven bases (DNA) to which the RNA polymerase binds.

(iii)RNA initiation site

The site where transcription begins is called initiation site. This is the region overlapping with the operator region.

Regular gene

The regulator gene directs the activity of the operator gene by producing inhibitor proteins called repressors. This repressor protein binds to the operator gene and blocks the path of RNA polymerase, thus preventing transcription. If an inducer is present in the system, it binds to the repressor which undergoes conformational change and becomes inactive. As the inactive repressor cannot bind to the operator, the structural genes get activated and protein synthesis continues.

In a repressible system, if only a repressor is present, it is inactive by itself and cannot attach to the operator gene and hence transcription continues. However, when another protein called co repressor is present, it attaches to the repressor forming a *repressor-corepressor* complex, which blocks the operator gene thus preventing transcription.

Types of regulatory mechanism

Earlier in the chapter it has been mentioned as to how the presence or absence of a substrate can induce or suppress the synthesis of the enzyme protein. Here we shall study in some more detail the various regulatory mechanisms that are involved in the synthesis of enzymes.

Induction. It has already been mentioned that in inducible systems, the inducible enzymes are normally present in small quantities and their synthesis can be increased manifold by the presence of the inducer. Such as example of inducer in *E.coli* is lactose and the inducible enzyme is β *galactosidase.* In the absence of lactose, only one ore two molecules of the enzymes β *galactosidase* may be present in the system. When the inducer is added, within three minutes about 3,000 molecules of the enzyme are synthesised by the gene.

When the inducer is absent, the repressor protein will block the production of the enzyme by not allowing transcription of mRNA by the structural genes. The repressor protein actually blocks the operator gene. When the inducer is present the active repressor binds to the inducer to form a repressor inducer complex. As a result, the repressor gets inactivated. An inactivated repressor cannot bind to the operator and hence the operator will allow the structural genes to transcribe mRNA and the enzyme is synthesised.

Repression. The presence of a metabolite inhibiting the synthesis of an enzyme in the system is caled repression. It has been mentioned already that histidine acts as a repressor and blocks the production of the enzyme responsible for histidine synthesis. This is called enzyme repression. Another example of repression is the blocking of the production of the enzyme tryptophan synthetase by the substrate tryptophan. The enzyme tryptophan synthetase is produced only when the bacterium (*E.coli*) is cultured in a tryptophan free medium.

In the repressible system, the protein produced by the regulator gene is an inactive repressor (*aporepressor*). On combining with a corepressor (effector), the aporepressor molecules get activated due to a conformational change. As a result of this the aporepressor becomes active and binds to the operator gene. As a result of the operator genes getting blocked the structural genes cannot by itself bind to the operator gene. The operator gene being free will allow the structural genes to transcribe mRNA. The following is a summary of the regulation of enzyme synthesis in inducible and repressive systems.

I. Inducible system

1. ***Inducer present :*** Regulator gene → Active repressor + inducer → repressor inactive → operator gene free → structural genes transcribe mRNA → synthesis of enzyme protein.
2. ***Inducer absent :*** Regulator gene → active repressor → binds to the operator gene which becomes inactive → no mRNA transcription → inhibition of enzyme synthesis.

II. Repressible system

1. ***Corepressor present :*** Regulator gene → inactive repressor + corepressor → active repressor → binds and inactivates operator gene → structural genes do not transcribe mRNA → inhibition of enzyme synthesis.
2. ***Corepressor absent :*** Regulator gene → inactive repressor → operator gene not blocked → structural genes transcribe mRNA → enzyme protein synthesised.

Negative control

In the inducible and repressive systems, protein synthesis is allowed to take place when the operator gene is free and it (protein synthesis) stops if the operator gene is blocked. In other words, gene expression is possible only when the operator is free. Such a controlled mechanism of protein synthesis is said to be of a negative type. A typical example of negative control is the production of the enzyme protein in the presence of lactose and in the absence of glucose. The presence of glucose actually inhibits the production of the enzyme β *galactosidase* while the presence of lactose promotes the production of the eyzyme.

In negative control, the regulator protein acting as the repressor prevents the transcription of mRNA by structural genes. The main controlling factor is the operator gene.

Positive control

In this control, the regulator protein functions as an activator and increases the enzyme synthesis. The activator binds itself to DNA in a location called the initiator site. Here the initiator site need not necessarily be the operator gene. As the regulator protein enhances the transcription of mRNA by structural genes this kind of regulation is called a positive control. A typical example of a positive control is the arabinose operon seen in *E. coli* (see later for details).

Replication of DNA is also an example of positive control. The bacterial genome is made up of replicating units called replicons which are capable of independent duplication. One of the structural genes of a replicon is known to produce an activator protein which in turn is induced by some cytoplasmic factors. The activator binds to a specific site on the replicon called the replicator and initiates replication from this point.

Catabolic repression

The works of Magasaine and others have shown that the presence of glucose inhibits the synthesis of enzymes involved in the breakdown of sugars such as lactose, arabinose or galactose. This phenomenon of repression is known as catabolic repression. This has been in many microorganisms including *E.coli*, yeasts etc. For instance, when a culture of *E.coli* maintained on lactose is supplemented with glucose, it has been seen that the synthesis of β *galactosidase* enzyme becomes suppressed as long as glucose is present in the medium. The inhibitory effect of glucose is due to a marked drop in the intercellular levels of a nucleotide cyclic AMP. The studies of Zubay and Beckwith (1970), have shown that glucose suppresses the level of cyclic AMP which in turn inhibits the transcription of mRNA.

In vitro studies of Eron, et al (1971), have indicated that lactose operon transcription requires not only cyclic AMP but also a protein called CAP (catabolic activator protein). CAP binds to the cyclic AMP. Thus, catabolic repression has two steps at the first stage CAP binds to cyclic AMP and at the second stage, this complex (CAP cyclic AMP) exerts its effect on the transcription process.

Feed back inhibition

In some instances, the final product of a particular biosynthetic pathway gets accumulated if it is not utilised for metabolic activities. Such a kind of accumulation many a time will cause the stoppage of further synthesis of the substance. For instance, if the normally growing *E.coli* cells are supplied with isoleucine (an amino acid), it has the effect of blocking the isoleucine synthesising enzyme. This inhibition is due tothe binding of isoleucine to enzyme *threonine deaminase.* This kind of inhibition is called feed back inhibition. Feed back inhibition may be defined as the inhibitory effect of the final product on the enzyme involved in the first step of production when a synthetic pathway is made up of several steps and is controlled by the activity of specific enzymes. Feed back inhibition has an advantage in the sense if the enzyme of the first step is blocked, there is no necessity to block the intermediary enzymes as in any case they would not get the substrate.

Feed back inhibition is mostly through the allosteric transformation of the enzyme so that its active site gets inactivated. Feed back inhibition is different from repression in that in the latter there is no mRNA synthesis, and hence no enzyme synthesis. Feed back inhibition is actually a case of control of gene expression after translation.

Role of Leader sequences in genetic regulation

Gene expression in prokaryotes is controlled both at the level of transcription as well as after the transcription. In the mechanisms which operate after transcription, a prominent role is played by leader sequences and attenuators. For example, in the tryptophan operon of *E.coli,* the mRNA of gene trpE (gene for anthranilate synthetase) has a sequence of 162 bases between initiator codon and the promoter operator region of this gene. This sequence of 162 bases is called leader sequence and is thought to exercise post transcriptional control. This leader sequence of the gene trp E has a mechanism which allows transcription when synthesis is needed, and stop the transcription whenever synthesis is not needed.

Regulation by antisense RNA

Some of the recent researches have shown that regulation of protein synthesis can be controlled at the translational level by using RNA which has complementary bases to mRNA. When such an RNA is introduced, it binds to mRNA forming RNA mRNA hybrid and prevents the translation, it (RNA) which is this RNA is complementary to mRNA and interferes with the translation, it (RNA) is called antisense RNA. Antisense RNA can be used in experimental conditions to stop the translation of mRNA. In *in vivo* conditions also, antisense RNAs are known to regulate the translation of mRNA. In transposon Tn 10, the production of the enzyme *transposase* is prohibited by antisense RNA. The production of antisense RNA is controlled by a promoter which directs transcription in a direction opposite to the transposase gene, thus producing antisense RNA.

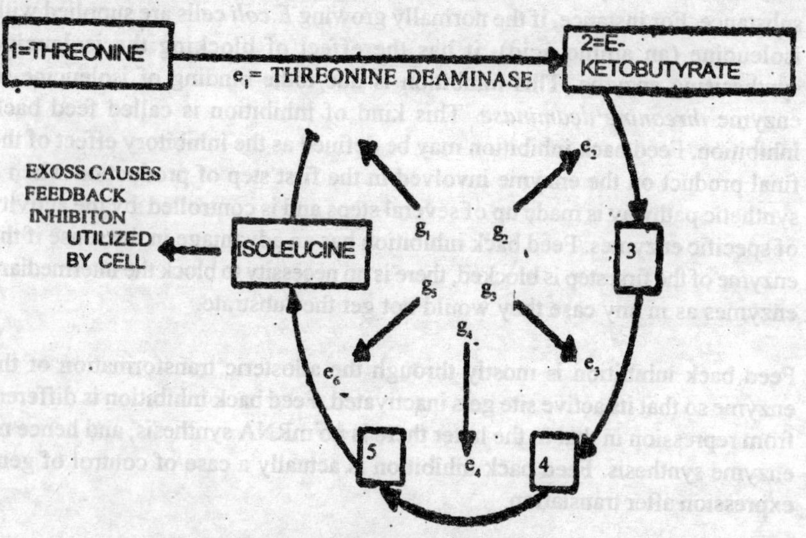

Fig. 8.2 Genetic regulation
Feedback inhibition in *E.coli* (after Stansfield, 1969)

Role of antisense RNA in regulating gene expression has also been demonstrated in the case of the two genes involved for the synthesis of outer membrane protein.

Examples of a few operon systems

The lac operon of E.coli. The lac operator consists of three structural genes.

1. Gene *z*. This is made up of 3,063 base pairs and codes for enzyme β *galactosidase,* which is active as a tetramer and breaks lactose into glucose and galactose.
2. Gene *y*. This is made up of 800 base pairs and codes for the enzyme β *galactoside permease* which is a membrane bound protein and helps in the transportation of metabolites.

3. Gene *ac*. This consists of 800 base pairs and codes for β *galactoside transacetylase*, an enzyme that transfers an acetyl group from acetyl coA to β *galactosides*.

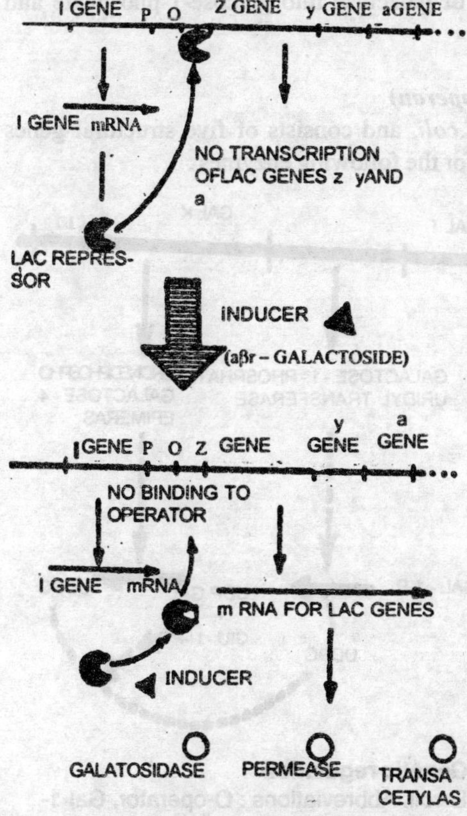

Fig. 8.3 Genetic regulation
Lac-operon model

In addition to the structural genes, the lac operon consists of a regulator gene *i*, an operator gene *o* and a promoter gene *p*. These six adjacent cistrons (genes) constitute the lac operon system. An operon can be defined as an integrated genetic unit.

In lac operon, the gene *i* produces a repressor (protein) which inactivates the operator by binding with it. Subsequently, an inducer such as lactose binds with the repressor molecule there by inactiviting it. As a result, the operon gene *o* is free from the repressor and becomes active. When the operator gene is active mRNA is produced from the structural genes *z, y* and *ac*. The production of mRNA starts from the promoter gene. It has been found that all the three structural genes produce a single mRNA molecule which codes for the sequence amino acids for all the three enzymes – β *galactosidase, permease* and *transacetylase.* Thus equal amounts of all the three enzymes are produced and the first message translated is for β *galactosidase;* this is followed by the synthesis of permease and then followed by transacetylase. Thus the operons show polarity. This has been proved by mutational studies. Mutations in the *z* region will reduce the quantities of all the three enzymes. While mutations in *y* locus will have no effect on β *galactosidase* but will reduce the quantities of other two enzymes. Such mutations are said to be **polar mutations.**

The gal operon system

In the gal operon system in *E.coli,* the sequence of three structural genes are gal E, gal T and gal K. These genes are directly responsible for the sequence of the enzymes produced namely galactokinase, galactose-1-phosphate uridyl transferase, and uridinediphospho-galactose-4- epimerase. These enzymes are responsible for the degradation of galactose into glucose-1-phosphate and UDP glucose.

Tryptophan operon system (trp operon)

The trp operon is also seen a *E.coli,* and consists of five structural genes ABCD and E. These genes code for the following enzymes :

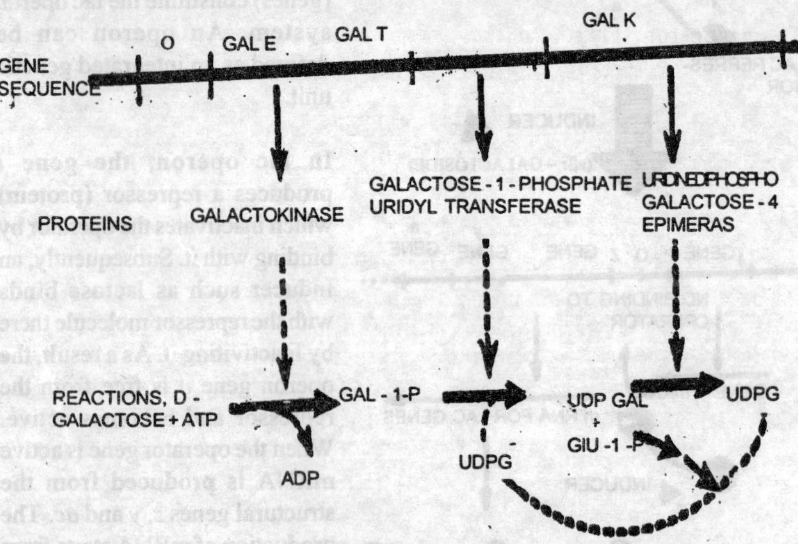

Fig. 8.4 Genetic regulation

The galctose (gal) operon of E.coli. Abbreviations : O-operator, Gal-1-galactose-1-phosphate, UDPG-uridine diphosphoglucose; UDP GA1-uridine diphosphogalactose; Glu-1-P=glucose-1-phosphage (after Goodenough and Levine, 1974)

1. Genes A and B code for tryptophan synthetase.
2. Gene C codes for PRA isomerase and IGP synthetase.
3. Gene D codes for phosphoribosyl pyrophosphate anthranilate transferase.
4. Gene E codes for anthranilate synthetase.

The levels of these enzymes is low when the amount of tryptophan is more and the level high when the amount of tryptophan is less. All these enzymes are regulated in a coordinated fashion. The operator and promoter are overlapping and are located next to *trp* E gene. The regulator gene (R) is located away from the operon and the repressor produced by it is inactive until it combines with tryptophan. All the enzymes are produced constitutively if the repressor is either absent or fails to combine with tryptophan. All the structural genes are translated together and form a polycistronic mRNA. This polycistronic mRNA besides the transcription of the bases of structural genes has a 160 nucleotide long leader sequence. This contains the attenuator site to which the substrate tryptophan can get attached and thus reduce the rate of transcription. Thus in *trp* operon, there are two regulation – the regulator gene and the attenuator site in the RNA itself.

The *trp* operon thus seems to have a dual control–a control by regulator and also by attenuator. In addition to this, feed back regulation is also seen in *trp* operon.

Arabinose operon

This is also seen in *E.coli* and is an example of positive control. Like the lac operon, the arabinose operon also consists of three strcutural genes. These are –

(i) *ara* A gene codes for arabinose isomerase
(ii) *ara* B gene codes for L-ribulokinase
(iii) *ara* B gene codes for L-ribulose-5-phosphate epimerase

These structural genes are located together on the *E.coli* chromosome. Their expression is under the regulation of a regulator gene C. This gene C called activator must be present for transcription of the arabinose operon. The product of C gene binds to the operator in the presence of arabinose and forms the activator protein. As a result of this, transcription takes place in the arabinose genes. If the activator is not bound to the arabinose or if the arabinose is absent, the operon is not expressed because the regulator protein binds to the operator gene (*ara* o) and functions as a repressor. In arabinose operon, the gene C acts both as an activator and repressor. In this way, arabinose operon is different from lac operon.

Gal genes in yeasts

In yeasts, galactose is fermented by the activity of three enzymes–kinase, transferase and epimerase. These are coded by the three structural genes GAL 1, GAL 7 and GAL 10. The regulator gene is located away from the structural genes. If a mutatkon occurs in a regulator gene, all the three enzymes are synthesised without control indicating thereby that the product of the regulator gene is a repressor. The regulator gene of gail in yeasts is very much similar to the gals in *E.coli*. A detailed genetic analysis has revealed that the product of the regulatory gene controls the functioning of an unlinked gene GAL 4. The product of GAL 4 is a positive regulator for the three linked genes. The GAL 4

gene has an operator gene *(c)* also. The regulator gene is *i*, the product of this gene exerts a negative control over gene c (operator). As a result, transcription of GAL 4 is reduced. When GAL 4 is free it produces a product which stimulates the transcription by the structural genes.

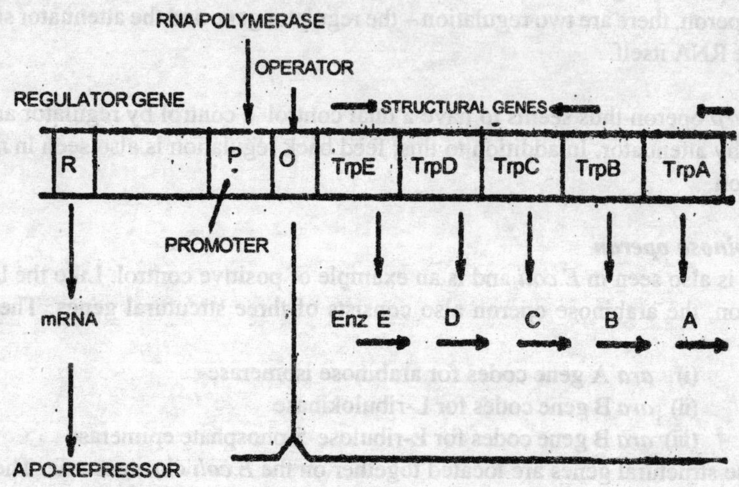

Fig. 8.5 Genetic regulation Tryptophan operon

Regulation of transcription in λ phage DNA

Regulation of the λ phage DNA has both negative and positive control mechanisms. The phage exists in a lytic or lysogenic state and this is controlled by some regulatory proteins. Depending on when the genes get activated during infection, multiplication or maturation phase, the λ genes can be classified into *early genes* and *late genes*. The early genes as the name indicates are required at the beginning of the infection cycle, while the late genes are required for the assembly and maturation of phages. The following table gives a list of genes identified on the λ genome along with their functions.

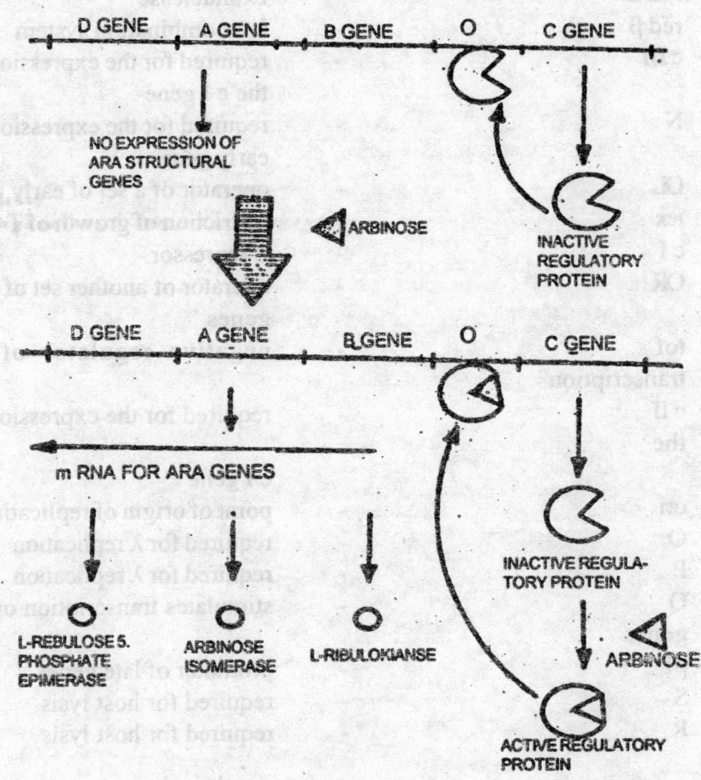

Fig. 8.6 Genetic regulation
The L-arabinose operon of E.coli

List of genes of the λ phage genome

A,W,B,C,D,E and F	-	Head genes
Z,U,V,G,T,H,L,K,Iand J	-	Tail genes
att λ	-	site of attachment and insertion
of the		
		phage DNA into the host DNA

int	-	integration of λ DNA e
xis	-	excision of λ DNA
red α	-	exonuclease
red β	-	λ recombination system
c III	-	required for the expression of the c I gene
N	-	required for the expression of early gene
OL	-	operator of a set of early genes
rex	-	restriction of growth of T4 R II
c I	-	λ repressor
OR	-	operator of another set of early genes
tof	-	negative regulator of c I transcription
c II	-	required for the expressions of the c I gene
ori	-	point of origin of replication
O	-	required for λ replication
P	-	required for λ replication
Q	-	stimulates transcription of late genes
pl	-	promoter of late genes
S	-	required for host lysis
R	-	required for host lysis

The cI gene codes for the repressor which binds to the operator genes thereby blocking the transcription of both the early operons. The repressor can bind to the operons blocking its integration and replication resulting in immunity due to a negative control exerted by a repressor. The product of gene N, however, acts as a positive regulator and allows transcription. The product of the N gene is known as antiterminator because instead of stopping, it allows transcription. Another example of positive regulation is the influence of Q gene product on late genes. This product combines with the coat transcriptase enzyme of *E.coli* and acts favourably on the promoter of the late genes. This is the λ phage, there are both negative and positive regulatory mechanisms to control the initiation and completion of lytic or lysogenic cycle. This is very necessary for an organism which has to survive in a continuously changing environment such as the bacterial cell.

GENETIC REGULATION IN EUKARYOTES

The mechanism of gene regulation of protein synthesis explained above illustrates the phenomenon in prokaryotes, whereby a repressor protein transcribed by regulator genes may activate or prevent the protein synthesis depending on the presence of inducer or corepressor.

In eukaryotes, genetic regulation takes place by a slightly different mechanism. Two types of proteins associated with the DNA of the chromosomes possibly play a role in gene regulation. These are histones and non-histone chromosomal proteins (NHC proteins). Since histones of different organisms have similar amino acid composition it is unlikely that they can selectively repress gene action. At best, histones may probably bring about a general (non specific) repression. It is believed that histones under certain situations bring about supercoiling of DNA preventing the movement of RNA-polymerase and thus blocking transcription. This is known by the fact RNA precursors are incorporated only in those regions where chromatin is diffuse (no super coiling).

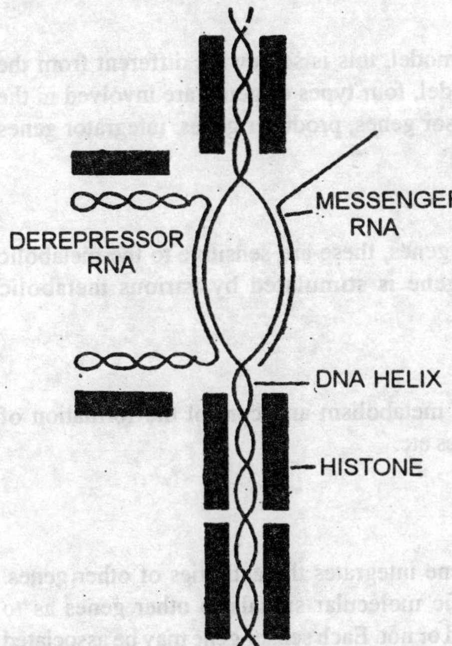

Fig. 8.7 Genetic regulation
Genetic regulation of protein synthesis in Eukaryotes-Frenster's model of gene regulation

Three models have been proposed by various scientists to account for gene regulation in eukaryotes. These are - 1. Frenster's model, 2. Non-histone depressor model and Britten and Davidson's model.

Frenster's model

This is also called gene specific derepressor RNA model. As has been pointed out earlier, histones bring about non-specific repression of transcription. Derepression (initiation of transcription) is brought about when nuclear polyions separate histones from DNA at certain specific parts (of DNA), and form complexes with histones.

Non-histones derepressors

The model proposed by Paul et al (1971), is similar to Frenster's model except the derepression is brought about by NHC acid proteins instead gene specific RNA. The fact that NHC proteins are of immense variety in different organisms, and also at different stages of growth even in the same individual goes to show that they can undertake selective derepression at specific gene loci. Derepression is brought about when NHC proteins bind themselves at specific loci to histones forming complexes. This complex separates itself from the DNA and allows it to transcribe, only at those specific loci.

Gene regulation by NHC proteins

(a) Non specific repression by histones.
(b) Specific binding of NHC proteins to histones.
(c) Removal of histone by NHC proteins.
(d) De repression of DNA, leading to transcription.

Britten Davidson Model

Also known as operon operator model, this is somewhat different from the other two models. As per this model, four types of genes are involved in the regulation process. These are sensor genes, producer genes, integrator genes and receptor genes.

1. Sensor genes

Comparable to the promoter genes, these are sensitive to the metabolic status of the cell. The sensor gene is stimulated by various metabolic substances.

2. Producer genes

These regulate the output of metabolism and control the formation of enzymes, cell organelle, membranes etc.

3. Integrator gene

As the name suggests, the gene integrates the activities of other genes. The integrator gene sends specific molecular signals to other genes as to whether the sensor gene is activated or not. Each sensor gene may be associated with a single or a group of integrator genes. In the latter case, a sensor gene may send a variety of molecular signals to other genes through its various integrator genes.

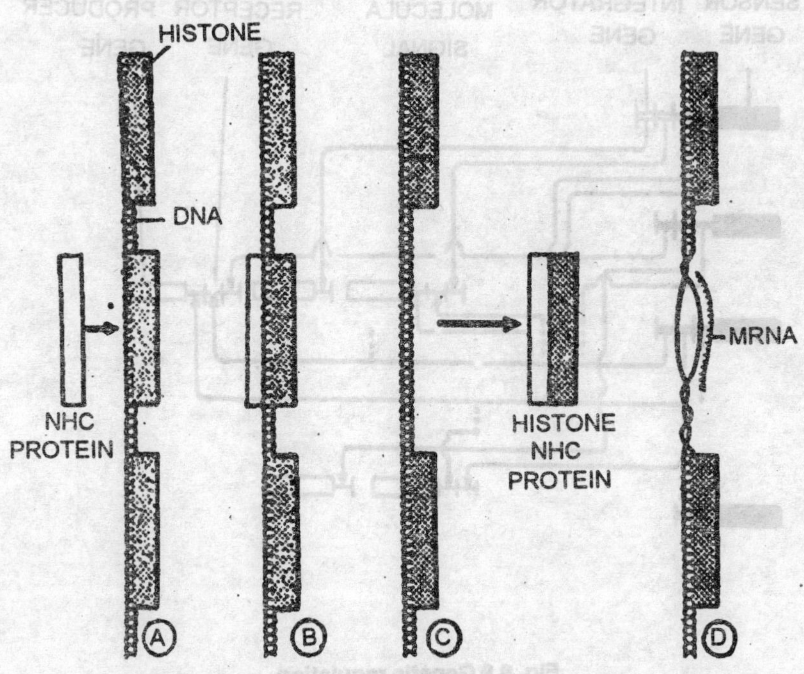

Fig. 8.8 Genetic regulation
Gene regulation by NHC histones

A. Non specific repression by histones
B. Specific bindng of NHC proteins to histones
C. Removal of histone by NHC proteins
D. De repression of DNA, leading to transcription

4. *Receptor gene*

This acts as an intermediate between the integrators and the producers. When the moleçular signal reaches the receptor gene, it inturn activates the producer gene. A producer may have one or a group of receptor genes associated with it in the same way as sensors have seveal integrators. This helps, not only in the variety of origin of molecular signals, (from sensors) but also variety in reception of these signals (by receptors). Perhaps, multiintegrator gene complex and multireceptor gene complex may have a specific multichannel relationship.

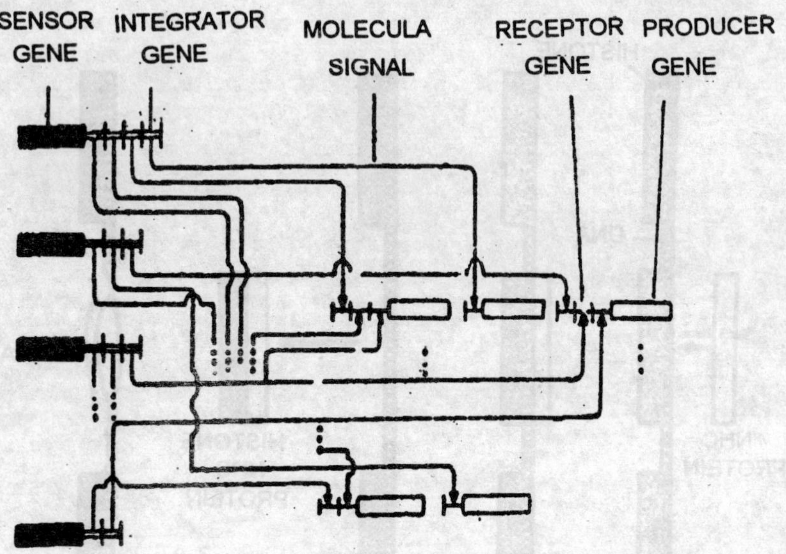

Fig. 8.9 Genetic regulation
Britten Davidson's Operon-Operator model

9

GENETIC RECOMBINATION

Genetical studies in microorganisms is extremely difficult, as they are very minute in size and do not possess any variability in their phenotype which can be subjected to genetic analysis. Another difficulty of microbial genetical studies is that microorganisms reproduce asexually to multiply their number and microbiologists thought that microbes like bacteria do not reproduce sexually. Extensive and repeated reseaches were made to explore the possibility of sexual reproduction in bacteria. Most of the earlier attempts were unsuccessful, because they were based purely on morphological observation of cell fusion. When Luria and Delburck established the assay procedure for obtaining the meaningful results, bacteriologists knew that their approach towards bacterial sexuality should change and a new technique is necessary for the analysis of sexuality in bacteria. The first successful experiment in bacterial genetics was performed by Joshua Lederberg and E.L. Tatum. These two collaborated in 1946, and using K_{12} strains of *E.coli* conducted what can now be called a classical experiment in bacterial crossing. They were able to demonstrate, that in *E.coli,* exchange of biochemical traits is possible. For their work on bacterial genetics, lederberg and Tatum shared with G.W. Beadle, the Nobel prize in medicine, for the year 1958.

With the establishment of the fact, that bacteria do exhibit sexuality, a vast area of microbial genetics opened up for geneticists. Prokaryotic organisms like bacteria were favoured for genetic experiments because –

1. Large amount of material could be handled as they are small.
2. Several generations could be obtained within a short period.
3. Pure cultures could be easily obtained for analysis.

With the advantages mentioned above, microbial genetics has made rapid strides in the last few decades

Types of genetic material transfer

As in any other organism in bacteria also, sexuality involves transfer and recombination of genetic material from one bacterial cell to the other. Three methods of genetic recombination have been demonstrated in bacteria. These are conjugation, transformation and transduction.

CONJUGATION

Genetic recombination

Lederberg and Tatum (1944), demonstrated this in experimenting with the biochemical mutants of *Escherichia coli.* These bacteria require the following four nutrients 1. Biotin 2. Methionine 3. Leucine 4. Thereonine. But they can be grown on a minimal medium (a medium containing only agar and carbohydrates) because, by using carbohydrates they can synthesize all the above four nutrients. They are called 'Paratrophs' and can be genotypically designated as follows :

$$B^+ M^+ L^+ T^+$$

In this, letter stands for the nutrient (B = biotin etc) and the sign + stands for the capacity to synthesize that particular nutrient.

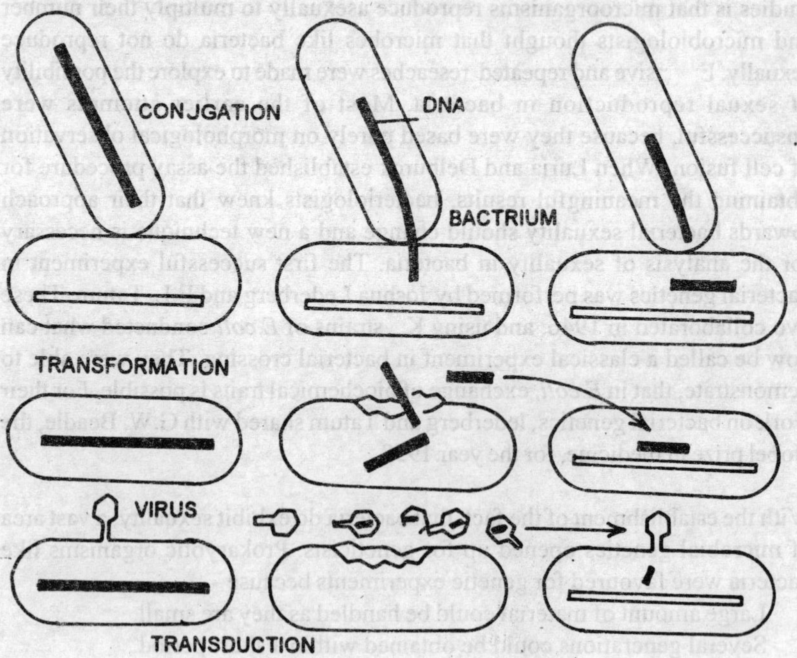

Fig. 9.1 Genetic Recombination
Genetic recombination in bacteria

By subculturing in the laboratory, the paratrophs undergo mutation and lose the capacity to synthesize one or more nutrients. The mutants are called *Auxotrophs*. By careful isolation, Lederberg and Tatum obtained two complementary polyauxotrophic varieties of *E.coli*. They can be designated as follows

1. $B^+ M^+ L^+ T^+$
2. $B^- M^- L^- T^-$

(The sign–stands for the loss of the synthesizing capacity. For ex : L means leucine cannot be synthesized, and hence it has to be added to the culture).

It is obvious that both strains 1 and 2 cannot be cultured on a minimal medium. Strain 1 cannot be cultured because it cannot synthesize leucine and theronine. Strain 2 cannot be cultured because it cannot synthesize biotin and methionine.

But when both of them are kept on a minimal medium, curiously they survive (normal expectation is, they should not live). On nutritional analysis, it was revealed that they were Paratrophs. The formation of paratrophs from two polyauxotrophic varieties can be explained only on the possibility of sexual recombination between strain 1 and 2.

$B^+M^+L^+T^+$ $B^-M^-L^-T^-$
1 2

Because of recombination of 1 and 2 the following combinations may be obtained.

$B^-M^-L^-T^-$ and $B^+M^+L^+T^+$
(a) (b)

Of the above, (a) will not survive because it has lost the capacity to synthesize all the nutrients, but (b) will survive because it has become a paratroph.

Mechanism of conjugation

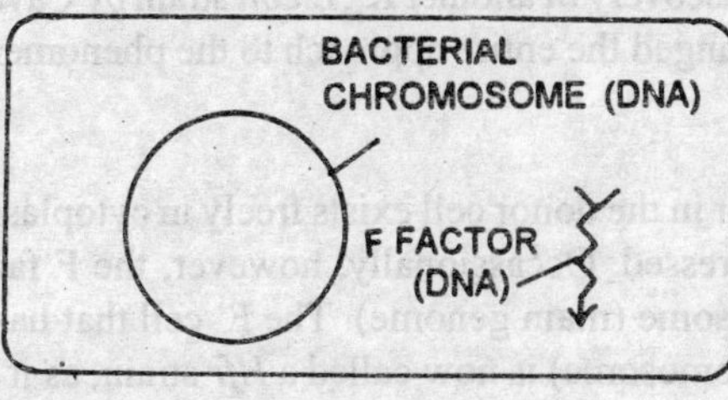

Fig. 9.2 Genetic Recombination
A bacterial cell showing circular DNA and F factor

The exact mechanism of genetic exchange was worked out by three groups of scientists led by Lederberg in Wisconsin USA, Hayes in London and Woolman in Paris. Genetic transformation was not involved in the process, was confirmed when cell free extracts of one bacterium failed to transformer other cells. It was demonstrated that the two bacterial cells physically come together and DNA from one passes into the other through the conjugation tube.

It was observed that in bacterial cultures, conjugation did not occur randomly between cells; rather it was selective. Genetic material seemed to migrate selectively only from particular cells to certain other specific cells. What is this specificity? What determines that only certain cells can conjugate while the others cannot. Careful studies indicated that the K_{12} *E.coli* has two strains – *Donor strain* and *recipient strain.* The donor strain possesses an episome (plasmid), while the recipient does not have this episome, as F+. The recipient strain does not have F factor and can be designated as F-. The donor strain may also be called male and the recipient female.

The fertility factor is free of the main genome of the bacterium and lies freely in cytoplasm. It is made up of a small DNA molecule having a molecular weight of 2,50,000. Normally, only one F factor is present in a cell and it replicates along with the main genome during cell division.

When conjugation occurs between F and F cells, the main genome is not involved. Consequently, chromosomal genes are not transferred from donor to recipient cell. In normal conjugation, the F factor of donor replicates and one copy of the F factor migrates along the conjugation tube to the recipient cell. As a result, both the cells now have a F factor. Consequent to this transfer, even the original recipient cell now becomes a donor as it has got an F factor. In other words, both the cells are now F^+. This means that the female cell has become male. The newly converted donor cell can now transfer its F factor to other recipient cells by *infecting* them with F factor.

If the entire conjugation process were to be limited to only transfer of the F factor with chromosome genes not being involved, it would have been perhaps of little genetic interest. But the discovery of another K_{12} *E.coli* strain by Cavalli-the *Hrf* (High fertility) strain changed the entire approach to the phenomenon of bacterial conjugation.

What is a *Hfr* strain? The F factor in the donor cell exists freely in cytoplasm – a point that has already been stressed. Occassionally, however, the F factor gets integrated with the chromosome (main genome). The F^+ cell that has an integrated F factor (with the chromosome) is now called a *Hfr* strain, as it can increase the rate of recombination many fold. So a F^+ male that brings about recombination has now been converted into a supermale (High fertility male). Integration of F factor into the chromosome involves a break (in the circular chromosome) and attachment of F factor. As a result of this integration, the chromosome along with its F factor becomes linear with two free ends. It is this chromosome. As the F factor becomes a part of the chromosome and as the chromosome has free ends, one end enters the other bacterium through the conjugation tube.

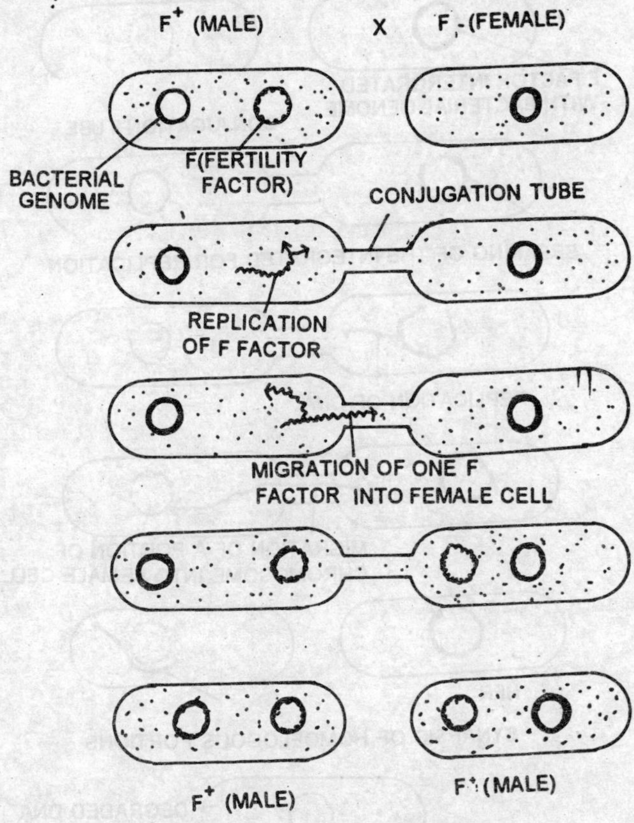

Fig. 9.3 Genetic Recombination

Conjugation between F+ (male) and F- (female) resulting in the conversion of female into male due to transfer of the fertility factor, F.

When a *Hfr* cell conjugates with a F cell, the latter receives the chromosomal genes. Which end of the chromosome of Hfr cell enters the F-cell? It has been found out that when F+ joins the chromosome with a single nick, it will be at one end. This end can be called posterior and the other end, anterior. Usually, chromosome movement from *Hfr* to F begins at the anterior end and the F factor enters the F strain only when conjugation is complete. This, however, rarely happens as conjugation and the F factor remains in the Hfr due to partial conjugation. Two possibilities exist here.

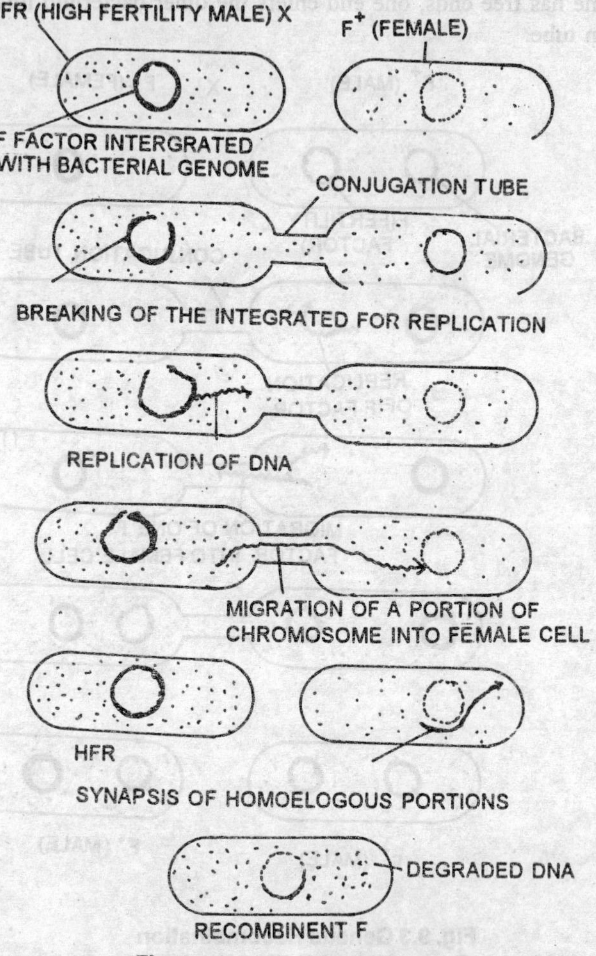

Fig. 9.4 Genetic Recombination
Conjugation between high fertility male (Hfr) and female (F-) cells, resulting in production of recombinant female

1. Complete conjugation

Total transfer of genome along with F factor from *Hfr* to F. As a result, not only the F acquires the genetic traits of *Hfr,* it becomes a Hfr. This is total recombination.

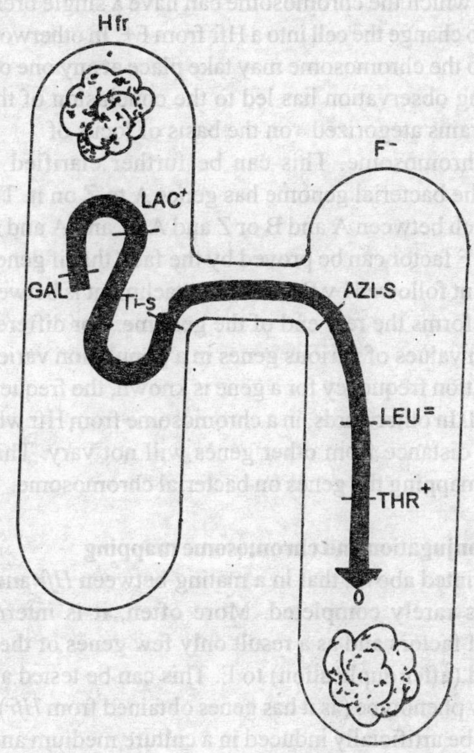

Fig. 9.5 Genetic Recombination
Transfer of genes from Hfr to F in a definite linear order

2. Interrupted or partial recombination

Here, chromosome entry from *Hfr* to F stops in the middle. As a result, F remains female (F) only, but acquires new chromosomal genes as a result of partial recombination. Whichever genes have entered from *Hfr* to F, they express themselves i.e., the F acquires new phenotype.

The F factor attachment to the chromosome is not at any fixed locus. There are several loci at which the chromosome can have a single break and the F factor can integrate to change the cell into a Hfr from F+. In otherwords, the integration of F factor into the chromosome may take place at any one of the possible loci. This interesting observation has led to the conclusion of the existence of several Hfr strains ategorized on the basis of locus of attachment of F factor to the chromosome. This can be further clarified as follows–let us suppose that the bacterial genome has genes A to Z on it. Theoretically, the F factor can attach between A and B or Z and A, C and A and soon. The point of attachment of F factor can be proved by the fact, that if gene A appears first in the recombinant followed by B, then the attachment is between A and B. The F factor always forms the rear end of the genome. For different Hfr strains, the recombination values of various genes in a population varies. But when once, the recombination frequency for a gene is known, the frequency of other genes is always fixed. In otherwords, in a chromosome from Hfr whichever gene may enter first, its distance from other genes will not vary. This aspect has been exploited for mapping the genes on bacterial chromosome.

Interrupted conjugation and chromosome mapping

It has been pointed above, that in a mating between *Hfr* and F, the process of conjugation is rarely completed. More often, it is interrupted by several environmental factors and as a result only few genes of the *Hfr* chromosome get transferred (after duplication) to F. This can be tested as the recombinant acquires a new phenotype (as it has genes obtained from *Hfr*). This interruption of mating can be artificially induced in a culture medium and the recombinant tested for acquiring the new trait. By carefully controlling the duration of mating one can observe, the series of new phenotype acquired by recombinant. Let us suppose, after 5 minutes of conjugation *gal* - gene enters the recombinant, and after 10 minutes *lac* - gene enters followed by leu-gene after 15 minutes, it is obvious that the genes are located in the following sequence.

gal- ———— lac- ———— leu-

By following the method of controlling conjugation time, one can map all the genes on a bacterial chromosome.

Jacob and Woman (1957), used the technique of "interrupted mating" between a *Hfr* strain and a F strain. At different time intervals, the mating mixture was diluted and agitated on a waring blender to separate the mating cells forcibly, thus interrupting conjugation and then plated on selective agar media. They demonstrated a lag period of 8 minutes before the first recombinant appeared. The recombinants then appear for other genes in a linear order in relation to the time allowed for conjugation. From this, Jacob and Wolman concluded that the latent period of 8 minutes is required for the conjugating cells to make contact,

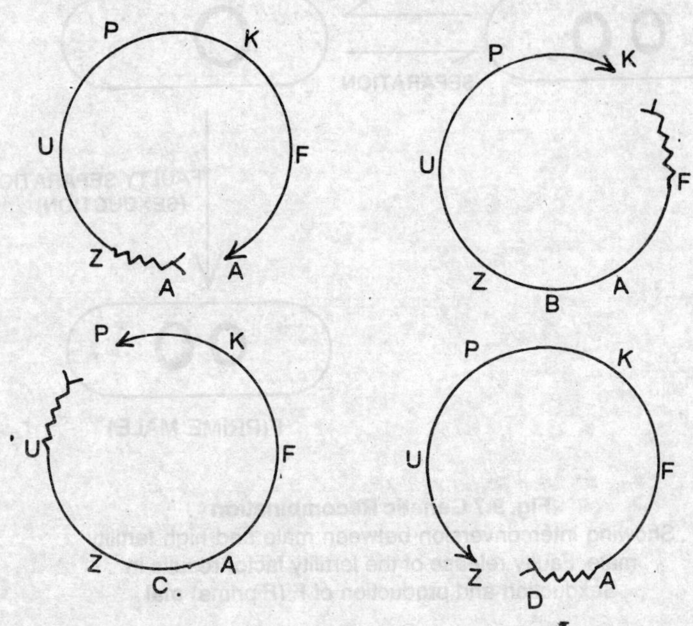

Fig. 9.6 Genetic Recombination
(A,B,C,D). Four hypothetical Hfr strains having different genes at their proximal ends

whereafter, the genetic material from *Hfr* is transferred to F in a linear order. They showed that 89 minutes are required for the complete transfer of the chromosome form *Hfr* to F. The bacterial chromosome is therefore, 89 time units long. Wolman and Jacob in 1956, produced the first genetic map of a bacterial chromosome in a *Hfr* strain.

The transfer of genes in linear order from male bacterium (tagged with virus) to female; the length of transferred segment is dependent the time taken for transfer, and the chromosome is, therefore, mapped in terms of time units.

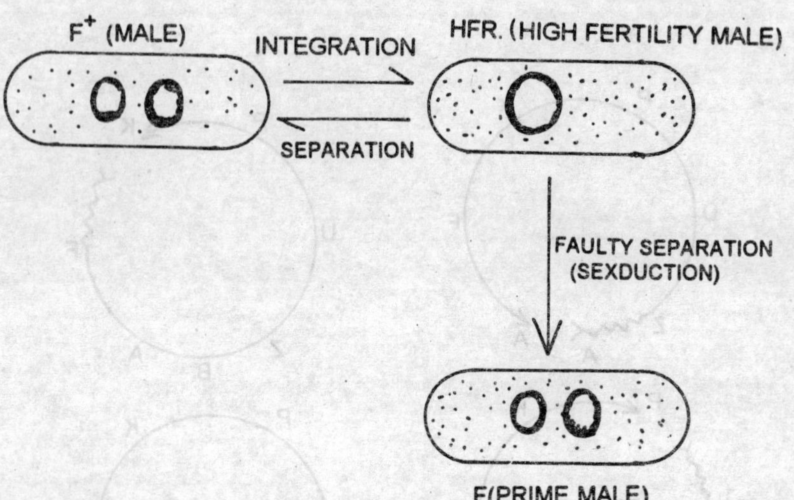

Fig. 9.7 Genetic Recombination
Showing interconversion between male and high fertility male. Faulty release of the fertility factor results in Sexduction and production of F (F prime) mal

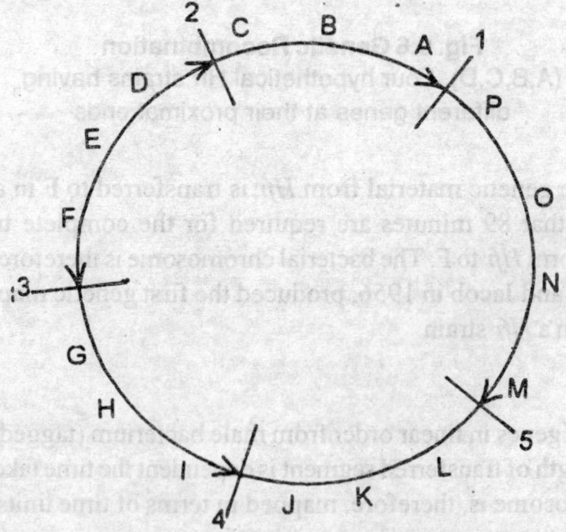

Fig. 9.8 Genetic Recombination
The linear order of genes in five hypothetical Hfr strains, resulting due to breakage of bacterial strain at different sites and subsequent orientation acquired by the resulting linear chromosome

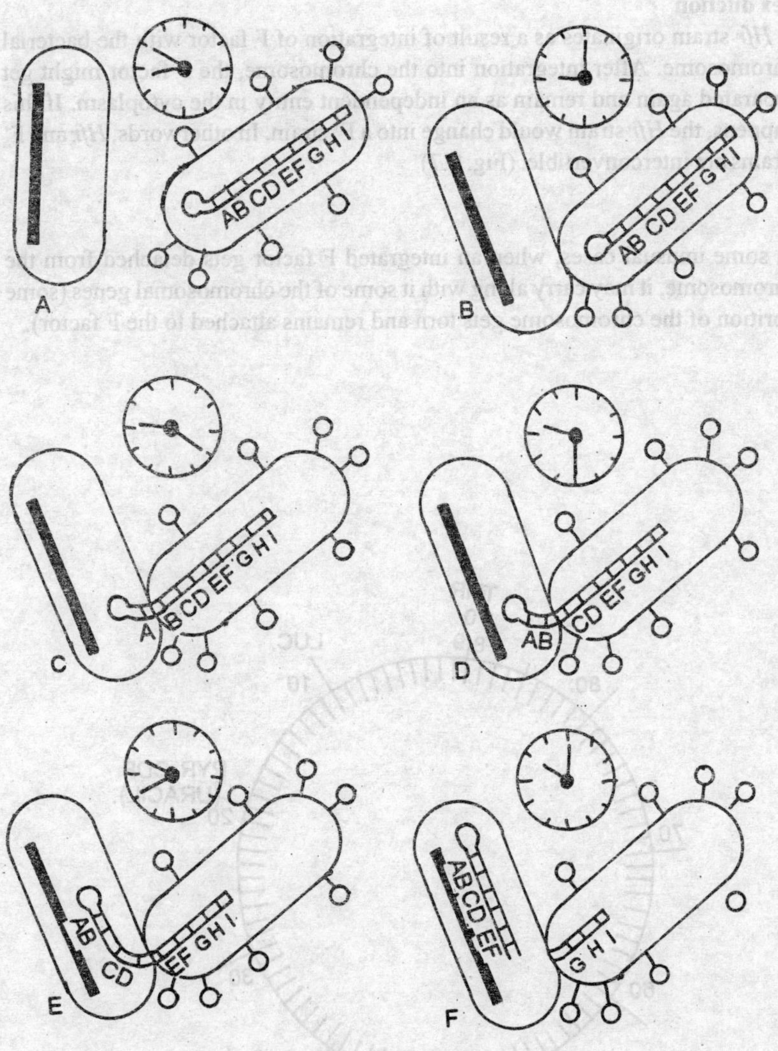

Fig. 9.9 Genetic Recombination

The transfer of genes in linear order from male bacterium (tagged with virus) to female; the length of transferred segment is dependent upon the time taken for transfer and the chromosome is, therefore, mapped in tems of time units

Sex duction

A *Hfr* strain originates as a result of integration of F factor with the bacterial chromosome. After integration into the chromosome, the F factor might get separated again and remain as an independent entity in the cytoplasm. If this happens, the *Hfr* strain would change into a F^+ strain. In otherwords, *Hfr* and F^+ strains are interconvertible. (Fig. 9.7)

In some unusual cases, when an integrated F factor gets detached from the chromosome, it may carry along with it some of the chromosomal genes (some porition of the chromosome gets torn and remains attached to the F factor).

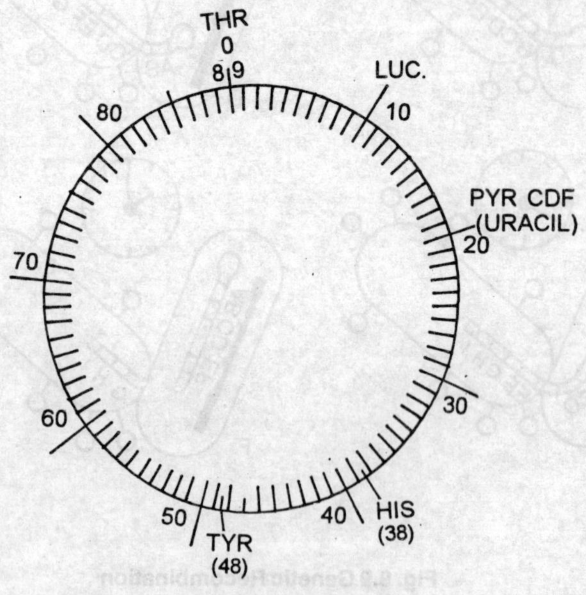

Fig. 9.10 Genetic Recombination
Circular linkage map of *Escherichia coli* chromosome, the units of recombination shown as minute required for transfer

Such an independent F^+ factor is given the name F^1 (F prime). An F^1 cell is male and like the F^+ except that the F factor has some genes of the chromosome.

When a F^1 (male) cell conjugates with a F cell, the recipient cell on getting the F^1 becomes a donor. In addition, it becomes diploid for certain genes of the chromosome which are present in the F^1. Because these genes are present on the chromosome as well as an F^1, this phenomenon is known as *sex duction* (Jacob and Alderberg, 1959). The recipient cells become *paritial diploids* for the genes introduced by the F^1.

GENETIC TRANSFORMATION

Every individual born out of sexual reproduction will have the mixture of traits of both the parents. In genetic terms, this means that there is genetic recombination in the off spring. Genetic recombination or transformation which is an index of genetic change has also been reported in bacteria.
Genetic transformation i.e., change of one genetic type to other was first reported by British bacteriologist F.Griffith (1928), while working with pneumococci (*Diplococcus pneumoniae*).

There are two strains in *Diplococcus pneumoniae.* One is capsulated, and the other one is non-capsulated. Non-capsulated ones form glistening smooth (S) colonies on the culture medium, while the non capsulated ones form dull, rought (R) colonies. Further, the capsulated (S) strains are virulent while the non-capsulated (R) strains are non virulent i.e., do not produce toxins. The ability of producing or not producing a capsule is a genetic character.
Griffith injected experimental mice with R strains of pneumococci and found that they survived, but when S strains are injected, the mice died. When heat killed (killing the bacteria by applying heat when DNA gets denatured and will not produce toxin) 'S' strains were injected the mice survived. Griffith next injected both R strain and heat killed 'S' strain to the same experimental mice and they (mice) died. Blood analysis of the dead mice (which were infected with R strain and heat killed S strain) showed the presence of 'live' S strains. There are three possibilities how the heat killed 'S' strains can become live again.

1. Heat killed S strain might have become alive in the body of mice–this is an impossibility.
2. A mutation might have occured in the R strain converting them to S strain. This is also not possible because mutations will not occur frequently, but everytime the experiment showed the presence of capsulated S strains in the blood of the dead mice.

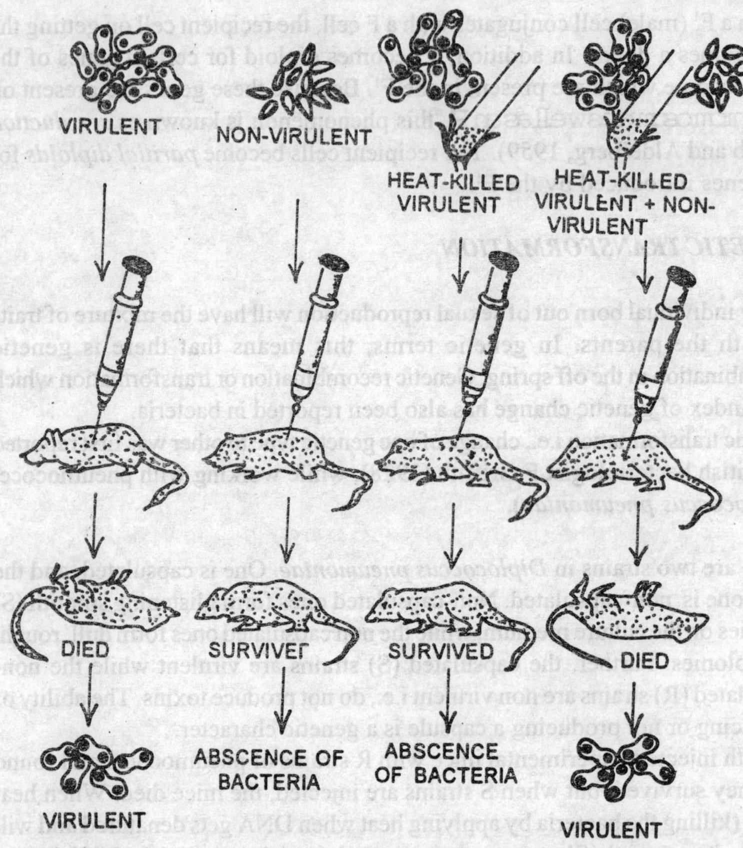

Fig. 9.11 Genetic Recombination
Schematic representation of Griffith's experiment to demonstrate transformation in *Diplococcus pneumoniae*

3. Some factor in the heat killed strain has transferred the genetic information for capsule synthesis to R strain making them produce the capsule. In other words, they become S strain. This is the only way a live S strain could be found in the blood of mice. Griffith (1928), thought that the transforming principle was a bacterial polysaccharide. But, Avery *et al* (1944), demonstrated that the transforming principle was DNA.

This experiment clearly demonstrates that a piece of DNA responsible for capsule synthesis recombined with the DNA of the R strain inducing it to produce a capsule. This can be called a natural method of genetic engineering.

Mechanism of transformation

This involves the following steps – competence, binding, penetration, synapsis and integration.

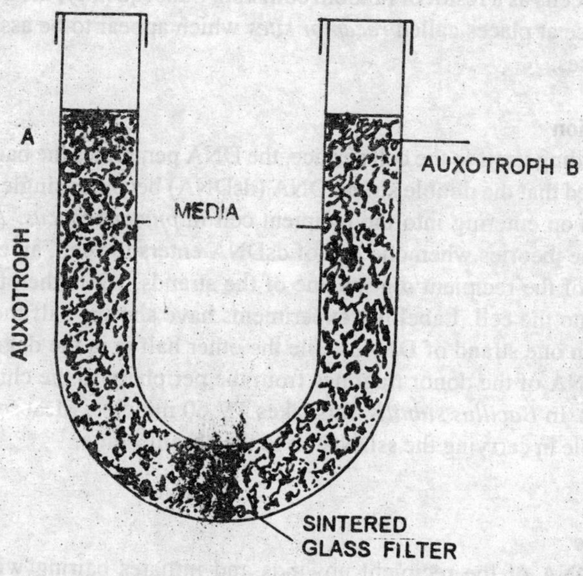

Fig. 9.12 Genetic Recombination
The Transduction (U tube) experiment of Zinder and Lederberg

Competence

The recipient bacteria must be competent if they are to accept the foreign DNA and undergo transformation. Competence of recipient bacteria depends on several factors such as age, physiological condition etc. The duration and time of occurrence of competence varies with species also. During growth, competence is observed during the end of logarthmic phase and just at the beginning of stationary phase. Experiments of Euphrissi–Taylor and Freed (1964), have shown that in *Pneumococcus,* competence developed in 90–100 minutes old culture and lasted for just 7–15 minutes. In *Bacillus subtilis* competence lasts for several hours, but only a few cells in the culture become competent. Several external factors are known to induce competence in an otherwise non competent cells.

Binding

DNA of the transforming bacterium comes into contact with the cell surface of recipient cells as a result of random collision. Subsequently, the DNA binds to the surface at places called *receptor sites* which appear to be associated with mesosomes.

Penetration

After binding itself to the cell surface, the DNA penetrates the bacterial cell. It is believed that the double strand DNA (dsDNA) becomes single strand DNA (ssDNA) on entering into the recipient cell in *pneumococcus.* According to one of the theories when one end of dsDNA enters the cell, an endonuclease enzyme of the recipient digests one of the strands, while the other strand is pulled into the cell. Labelling experiments have shown, half the labelling is present in one strand of DNA, while the other half is in the degraded strand. The ssDNA of the donor migrates from the periphery to the chromosome of recipient. In *Bacillus subtilis,* this takes 15–60 minutes. Mesosomes seem to play a role in carrying the ssDNA on to the chromosome.

Synapsis

The dsDNA of the recipient unwinds and initiates pairing with the donor ssDNA. The donor ssDNA is identical, but not exactly homologous with the recipient DNA. This is necessary as otherwise there would be no variation.

Integration

The unpaired recipient DNA strand is cut in two places by an endonuclease enzyme. An exonuclease enzyme also degrades the unpaired portion of ssDNA. Integration of donor DNA is brought about when *repair synthesis* of DNA takes place to fill any gaps. These are joined by ligase. The transformed heteroduplex undergoes replication to produce copies when the recipient cell divides and expresses its phenotype ie., virulence in the case of *Pneumococci.*

TRANSDUCTION

Transfer of genetic material from one bacterium to another mediated by bacteriophages is known as transduction. Transduction was first discovered by Zinder and Lederberg (1952), in *Salmonella typhimurium*, and subsequently, by other workers in *Escherichia, Shigella, Staphylococcus* etc.

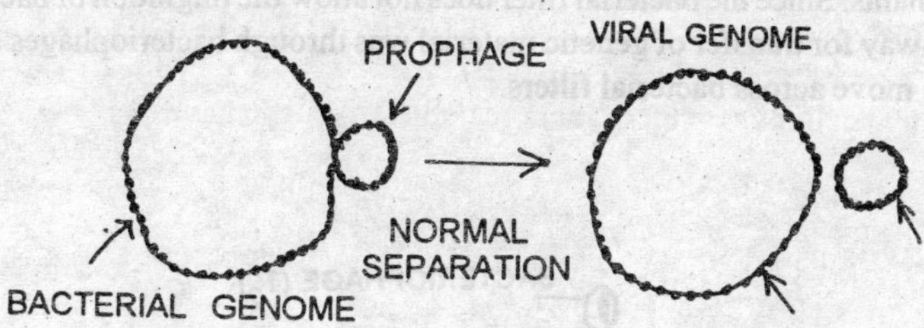

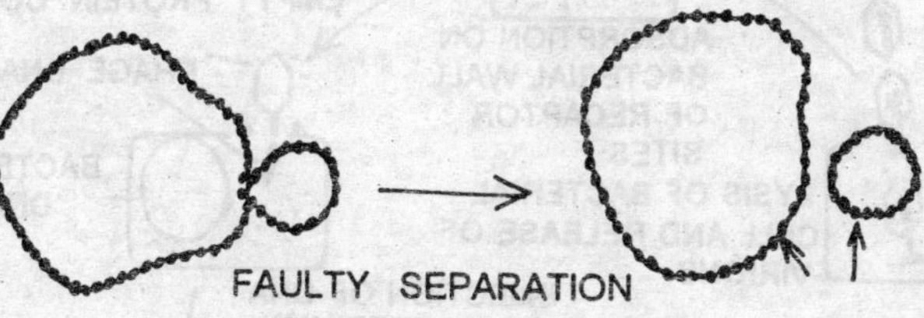

EXCHANGE OF BACTERIAL AND VIRAL GENOMES

Fig. 9.13 Genetic Recombination
Showing normal and faulty separation of phage DNA from bacterial genome. A piece of bacterial DNA is incorporated into the phage DNA and vice-versa.

In transduction, when bacteriophages infect a bacterium they often incorporate a portion of the genetic material of bacterium into their genetic system. If such a bacteriophage infects another bacterium (of the same species), the bacterium genes (of the first infection) may get incorporated into these bacteria. These bacteria now exhibit some of the traits of the first strain. This shows that bacteriophages have picked up some genes from the first bacterium and transferred it on to the second bacterium.

In order to prove bacteriophage mediated genetic transfer. Zinder and Lederberg performed a 'U' tube experiment. The experimental set up consists of a 'U' tube where the two arms are separated at the base by means of a sintered glass filter (Bacterial filter), Zinder and Lederberg (1952), grew two different strains of bacteria *Salmonella typhimurium* in the two arms. One strain was infected with bacteriophages. A little later, when the bacteria in both the arms of the 'U' tube were analysed, it was noticed that the strains in both the arms had become recombinants. Since the bacterial filter does not allow the migration of bacteria, the only way for transfer of genetic material was through bacteriophages since they can move across bacterial filters.

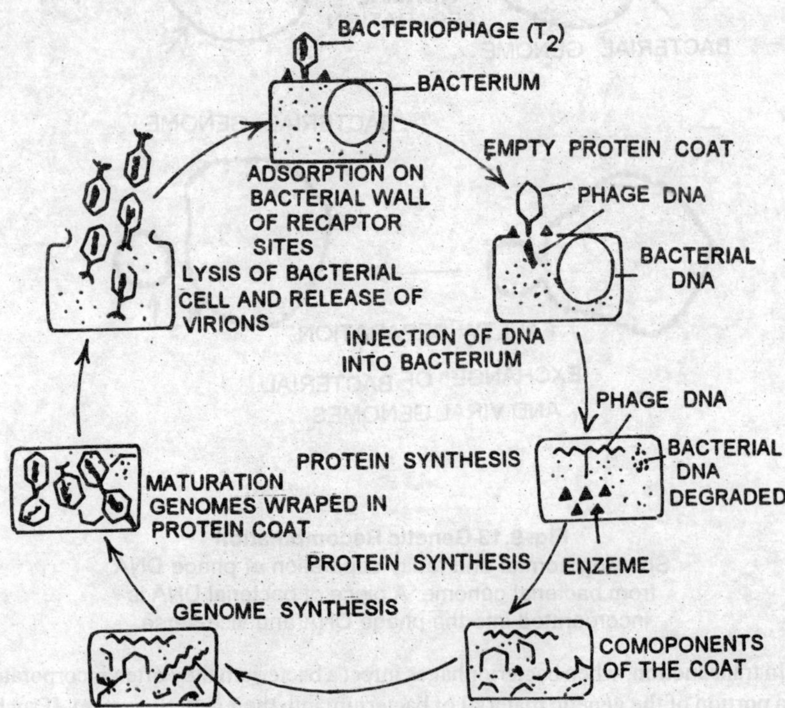

Fig. 9.14 Genetic Recombination
Lytic cycle of a bacteriophage (T_2).

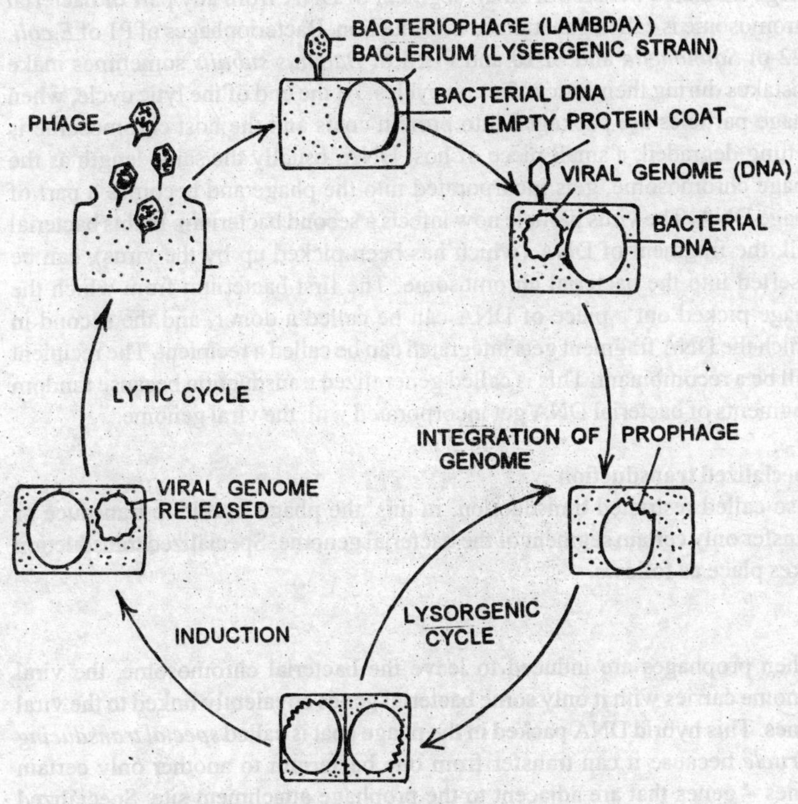

Fig. 9.15 Genetic Recombination
Lysogenic cycle of λ bacteriophage

Types of transduction

There are two types of transduction – Generalized transduction and special or restricted transduction.

Generalized transduction

Phage mediated transfer of small segment of DNA from any part of bacterial chromosome is called generalized transduction. Bacteriophages of P1 of *E.coli*, P22 of *Salmonella* and SP10 and PBSI of *Bacillus subtilis* sometimes make mistakes during their lytic infection cycles. At the end of the lytic cycle, when phage particles are packaged into protein coats and the host chromosome is getting degraded, a small piece of host DNA, usually the same length as the phage chromosome, gets incorporated into the phage and becomes a part of phage DNA. The virus particle now infects a second bacterium. In this bacterial cell, the fragment of DNA (which has been picked up by the virus), can be inserted into the bacterial chromosome. The first bacterium from which the phage picked out a piece of DNA can be called a donor, and the second in which the DNA fragment gets integrated can be called a recipient. The recipient will be a recombinant. This is called generalized transduction because random fragments of bacterial DNA get incorporated with the viral genome.

Specialized transduction

Also called restricted transduction, in this, the phage is able to transduce or transfer only certain segment of the bacterial genome. Specialized transduction takes place as follows.

When prophages are induced to leave the bacterial chromosome, the viral genome carries with it only some bacterial genes covalently linked to the viral genes. This hybrid DNA packed in the phage coat is called *special transducing particle* because it can transfer from one bacterium to another only certain genes – genes that are adjacent to the prophage attachment site. Specialized transduction is induced by λ 80 prophage, which usually occupy a fixed site on the bacterial chromosome. The site is called prophage site. In *E.coli,* the prophage site is flanked on one side by bacterial genes gal K, T and ε genes of the galactose gene cluster, and on the other side by block of genes involved with the synthesis of biotin–bio A,B,F,C and D. Thus, λ prophage can transduce only *gal* or *bio* genes.

The total length of hybrid DNA in a transducing prophage is exactly equivalent in length to the normal viral genome (as otherwise it cannot be packed into the coat), which means to accomodate the bacterial genes, a corresponding length of viral genome should be missing. As such, these phages are called defective paste (they lack some viral genes).

Abortive transduction
In this, the foreign DNA (of the donor tagged to a viral genome) does not get incorporated into the genome of recipient bacterium. The DNA fragment remains free in the cytoplasm and cannot undergo replication. When the bacterial cell divides, only one cell in each generation will have the foreign DNA (exogenote), while all the others are non-transducted. The non -transduced cells might have the gene product (produced by the exogenote before cell division), but progressively it gets reduced as it is distributed among the cells. This phenomenon which brings about incomplete transfer of genes is called *abortive* or *incomplete transduction.*

Growth phases of transducing phages
Bacteriophages which a bring about transduction can be divided into two groups based on their growth cycles. These are **lytic cycle** and **lysogenic cycle.**

Lytic cycle
This is seen in phages called virulent phages typified by T series phages. Here, soon after infection, the host cell lyses releasing the newly formed viral particles. The lytic cycle seen in Tphages infecting *E.coli* bacterium has the following stages.

Adsorption
The phage attaches to specific receptor sites on the bacterial surface.
Injection
The tail fibres hold the phage in position and a pore is formed in the bacterial wall, through which the viral DNA is injected inside the host cell.

Protein synthesis
Phage DNAm 'takes over' the synthetic machinery of the bacterial cell, and forms viral mRNA and through this the viral proteins.
Genome synthesis
Viral DNA replicates and produces several copies of the viral genome.

Maturation and release
Viral genome gets packed into the newly formed coats, thus assembling new phage particles. At a particular time, the host cell lyses (breaks up) releasing the phage particles.

Lysogenic cycle
This is seen in λ phage. The phages are also called temperate phages (as they show lysogenic cycle). Lysogenic cycle was first discovered by Andre Lwoff.

The characteristic feature of this cycle is that the host cell do not lyse as a result of infection by the phages. But the phages will remain in the host cell in a sort of an association. The lysogenic cycle is as follows.

The DNA of the phage, after adsorption and injection, enters the bacterial cell. But, instead of starting its degrading activity, the phage genome gets integrated with the host genome. This integrated condition of the viral genome is said to be **prophage.** When the bacterial cell divides, the integrated genome replicates in the normal way. The viral genome may be said to be in a hibernating state as it is not trying to express itself. Occasionally, the viral genome gets separated from the bacterial genome and is released into the cytoplasm where it starts multiplying and causes the lysis of cells. The dissociation of viral genome from bacterial genome (with, causes lysogeny to become lytic), is called ***induction.***

10

GENE MUTATION

Evolution in the biological world is primarily dependent on two events to bring about hereditary variations. These are - **mutation and recombination;** while mutation provides the raw material for the changed potentiality of inheritance, recombination permits permutation combinations in the gene pool, thus stabilizing the mutational change in the environment.

What are mutations? To understand this we will briefly analyse how characters are trasmitted from parents to offspring. We know that the characters are trasmitted from parents to offspring through the genes. But the offspring will not have the same genes as those of the parents, rather they have copies of them. In other words during reproduction, parental genes produce copies of themselves to be passed on to the offspring. Gene copying or gene reproduction is a precise process and generally there are no mistakes. Occasionally however, mistakes may occur while copying, so that the newly formed gene gets an altered structure. This modified gene then goes on multiplying and produces its own copies just as the original gene does. This alteration in the changed gene structure is called **gene mutation.**

Broadly speaking, mutation includes all those phenotypic variations, which arise due to altered gene structure and are heritable. Hence the term encompasses, small and minute changes brought about by altered individual genes (gene mutation) and large changes brought about by the altered chromosome structure or number. These latter types of changes can also be called chromosome mutations, and they have been described under the chapter chromosomal aberrations. In this chapter we will study gene mutations. Gene mutations are also called point mutations as they occur at a specific gene locus altering its molecular structure. Sometimes chromosomal aberrations also can bring about a slight phenotypic change, in which case it is difficult to distinguish between these two.

History : The term mutation was first coined by Hugo devries in the early 1900 to designate sudden and drastic changes that occured in plants. Devries was conducting breeding experiments on the plant *Oenothera lamarckiana* (of the family Onagraceae) commonly called the *Evening primrose*. He noticed that some of the progeny possessed characters which were not to be found in either of the parental types. He further observed these changed types bred true. He therefore described these sudden changes as mutations and ascribed them to some changes occuring in the germplasm of parents.

Even Charles Darwin was aware of mutations, even though he did not use the term (mhtations). He was aware of sudden changes occuring in organisms both plants and animals and had used the term sports, to describe them.

The first recorded instance of mutation dates back to 1791, to the pen of Seth Wright, a New England farmer, where appeared a male lamb with short bowed legs. Wright reared this lamb and it bred true. A lamb with short legs was of great advantage as it could not jump even a low level fence. This short legged breed was called **Ancon breed.** This breed however became extinct after about eight years. But the same Ancon sheep appeared some fifty years later in the flock of a Norwegian farmer, representing probably the repetition of the same mutation. Since then a number of mutant types have been discovered. Hornless individuals in cattle, double toed cats, albino rats, mule footed swine, dwarf cupid sweet pea, double flowered, white flowered varities of many plants are all mutants.

The scientific study of mutations began with the work of Morgan on *Drosophila* in 1910. He observed that among a population of red eyed flies, some individuals with white eye appeared and they were predominant among the male individuals. He further found out that the white eyed gene is recessive to red eye and located on the X chromosome. Since the male individuals possessed only one X chromosome and the females had two X chromosomes, the recessive character appeared more frequently among males than among females.

After the original discovery of the white eyed mutant in *Drosophila* by Morgan and his co-workers, a detailed analysis of millions of these fruit flies for over 17 years was carried out by different geneticists and nearly 500 gene mutations have been discovered. Simultaneously other organisms like maize, rodents, pea, snap dragons, poultry, man etc. also have been investigated and scores of mutations have been discovered. Even in microorganisms also mutations have been detected.

Size or range of mutations : Size of mutation varies in different organisms and in different genes. These may be very minute and may not be detectable at all or as in some cases (Ancon breed), so striking that they can be easily made out. In some microorganisms like Bacteria, Fungi *(Neurospora),* Blue green Algae etc, mutations may not bring out any morphological effect but may bring about some changes in their nutritional requirements. In some cases mutations may be so large as to even result in lethality to the individual.

Fig. 10.1 Gene Mutation
Ancon sheep with ram and ewe

Frequency of mutations : The frequency is generally less unless there is a drastic change of the surroundings. According to Muller, in *Drosophila,* the mutation rate for any gene is less than one in one million. It has also been calculated that a particular gene in *Drosophila* may mutate only once in every 40,000 years.

The rate of mutation also varies between different individuals, organisms and sometimes even under different ages of the same organism. J.B.S. Haldane, a renowned geneticist has estimated that in a comparison between man and *Drosophila,* the former has less chances of mutation as most of them are harmful so there is a selection pressure in man against mutations. If the mutations in man were to take place at the same rate as in *Drosophila,* the whole genetic system of the human race would be impaired. As against the unaltered life expectancy of 40,000 years for a *Drosophila* gene, the human gene has a life expectacy of 25,000,00. In one of the case studies of Chondrodystrophic dwarfness (in these human beings, arms and legs are short, while other parts are normal), it was estimated that one in every 11,500 births haye this genetic defect (the data obtained from a Denmark hospital). This gives a rate of one mutation for every 23,000 genes at this locus. If we assume, the average age of the parents to be 30, who give birth to dwarf children, then one can estimate that the gene mutates once every 6,900,00 years. The following two tables give the frequency of mutations in man and Corn.

Table : Rates of mutation in some human genes

Sl. No	Phenotypic effect	Genetic condition	Frequency in %	Occurrence oncein every gametes
1.	Colour cannot be distinguished	Colour blindness	0.0028	35700
2.	Blood clots very slowly	Hemophilia	0.0032	31250
3.	No pigmentation in hair, skin etc.	Albinism idiocy	0.0028	35700
4.	Mental faculties dull after birth	Amaurotic	0.0011	90909
5.	Limbs and arms dwarf	Chondrodystrophic dwarfism	0.0070	14300
6.	Face with red patches, tumours in Kidney, brains etc., at later stages	Epiloia	0.0012	83333
7.	Iris in the eyes absent	Aniridia	0.0005	2,00,000
8.	Abnormal WBC resulting in reduced immunity	anomaly	0.0080	12,500

Table : Rates of mutation in maize (After Stadler 1943)

Sl. No.	Gene	Phenotype	Frequency (1 in one million gametes)
1.	R	Plant colour	492
2.	Cl	Chlorophyll	106
3.	Dr	Red aleurone	11
4.	WX	Normal endosperm	0
5.	Su	Sugary endosperm	2
6.	Sh	Full endosperm (not shrunken)	1
7.	Ci	Aleurone colour inhibitor	2
8.	Y	Yellow endosperm	2

Mutational Unit : Earlier it was not known as to what constitutes a mutation in a gene. But recent researches have shown that the smallest amount of DNA within which mutation can take place is called a *muton.* At the molecular level mutation involves alteration in the base pair sequence.

Mutator genes and mutable genes : It is difficult to assess (among plants and animals) as to which gene frequently mutates and which gene is comparatively stable. But in some of the orgamisms like maize (Emerson), it has been observed that some genes change quite frequently, w-hile others are not so. Barbara McClintock (Nobel laureate, discovery of jumping genes) has shown in maize that some genes are not only stable, but influence others to mutate. The genes that influence others (to mutate) are called mutator genes and the genes that are so influenced are called mutable genes. These genes, mutator and mutable are more frequent in plants than in animals.

Classification of mutations : Broadly speaking, mutations are of two types - Chromosomal and gene. Chromosomal mutations actually do not change the gene structure, they bring about change in the phenotype either due to loss, addition, altered seqeuence or trasferred position of the genes. Gene mutation however changes the phenotype by basically altering the gene structure itself at the molecular level.

Gene mutations are basically of two categories - **Spontaneous and induced.** Spontaneous mutations apparently arise without the help of any identifiable external agency. They arise due to some internal changes. Induced mutations on the other hand are traceable to some external agency like radiation, temperature, chemicals, gas etc.

Mutations may also be further classified into various categories based on the effect, cell type inwhich it occurs, direction of mutation etc. The following is a brief classification of such mutations.

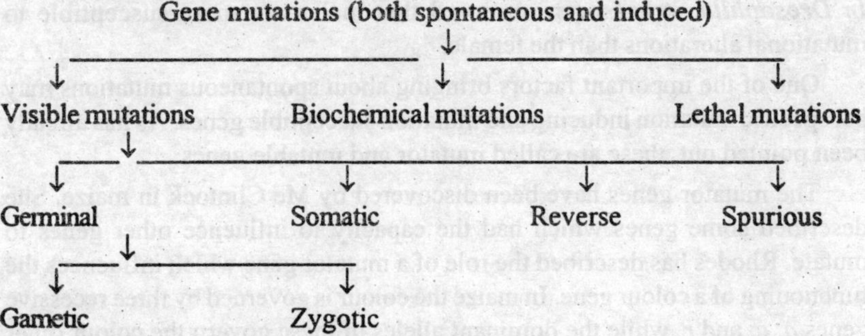

We shall briefly study a few examples of spontaneous and induced mutations and then study a few instances of the specific types of mutations above.

SPONTANEOUS MUTATIONS

As has already been pointed out, the agency which brings about spontaneous mutations is not known. They apparently occur due to natural causes. The first recorded instance of a spontaneous mutation is that of the Ancon breed sheep. A detailed study of the spontaneous mutation in *Drosophila melanogaster,* is that of T.H. Morgan, where he described the sudden appearance of white eyed flies in a population of red eyed individuals. Some of the spontaneous mutations noticed in plants and animals are as follows.

Spontaneous mutations in plants : 1) Double petunia, 2)Double rose, 3) Shirley poppy, 4) Single flower mutant in sunflower, 5) Multileaf formation in tobacco reported by Hayes, 6) Reduction in ray size in the ray florets in *Bidens pilosa* of Compositae(reported by the the author).

Spontaneous mutations in Animals : 1) Polydactyly in man, sheep, cats etc. 2) Multinippled condition in sheep, 3) Silky fowls, 4) Albinism in man, pig, pigeon, rodents etc. 5) Double cared cattle, 6) Mule footed swine, 7) White canaries, 8) Hornless cattle, 9) Hares' lip in man, 10) Widows peak in man, II) Lobster claws in man, 12) Cleft palate in man etc.

Spontaneous mutations are generally rare and may be of two types dominant and recessive. Usually recessive mutations are more frequent than dominant mutations.

Factors affecting spontaneous mutations : The rate of spontaneous mutations are known to change under the influence of various external and internal factors. One of the external factors which might bring about spontaneous mutations is a change in the temperature range (normally a higher temperature makes the genes susceptible for change). Disturbed metabolic rates may also bring about mutational changes. Age and sex may also have a bearing on the mutational rate. Increased age may render the genes prone for mutational change. *In Drosophila,* it has been observed that males are more susceptible to mutational alterations than the female.

One of the important factors bringing about spontaneous mutations may be, specific mutation inducing and mutation succeptible genes. As has already been pointed out, these are called mutator and mutable genes.

The mutator genes have been discovered by Me Clintock in maize. She described some genes which had the capacity to influence other genes to mutate. Rhodes has described the role of a mutator gene which influences the functioning of a colour gene. In maize the colour is governed by three recessive genes *a, c,* and *r,* while the dominant alleles of these govern the colour other than green. In addition to these there is a mutator gene)((dotted colour) which controls the variegated colour in the grain kernel. This gene influences the gene *a* to mutate so that red patches are produced. Usually the plants with the genotype *aa cc RR* will be green (all the three dominant gene pairs are required to produce colour other than green), but the presence of *Dt* makes it variegated.

Mutable genes other than the one mentioned above (a) have been reported in maize by Emerson. He investigated a variety of maize withwhite pericarp having several red spots of varying sizes; the homozygous recessive allele producing the white colour. According to Emerson, these red patches are produced by the unstable white gene undergoing spontaneous mutation. Each red spot in the endosperm indicates an area where a mutation has occurred. Jones further continued the studies of Emerson and found that there are six red spots, each having 10 - 45 coloured cells. These coloured cells were exactly identical with those produced by the pure red gene. Jones was conducting an experiment with a homozygous plant and he beleived that the six red spots indicated that the gene for white had mutated six times producing a red cell every time. These cells further multiplied and produced the red patches. The gene for white endosperm colour is thus a typical example of a mutable gene.

Several such mutable genes have been reported *in Drosophila,* sweeet pea, 4 o' clock plant, snapdragons etc. The genes for miniature wings, gene for purple eye colour in *Drosophila* are known to be mutable genes. In maize plant itself, Stadler has observed that the gene R for aleuron colour mutated some 492 times in about a million gametes that he studied.

INDUCED MUTATIONS

Various external agencies are known to bring about mutational changes in genes. All these are included under the general category of induced mutations. H.J. Muller (1927) may be regarded as pioneer in radiation genetics. In his studies on *Drosophila,* he showed that flies subjected to X radiation undergo mutations. Similarly Stadler (1928) demonstraded the mutational effects of X rays in barley and maize. Various kinds of agents causing mutational changes in organisms are known. These are of the following types, a) Radiation b) Temperature changes and c) Chemicals.

Radiation induced mutations : Various types of radiation (excluding visible light) are known to cause mutations in different intensity. Radiations are of two categories depending upon their capacity of penetration into a substance -*Ionizing* and *non ionizing.* Generally speaking, non ionizing radiations are not of a high potentiality to bring about mutations at a rate which is brought about by ionizing radiation. The Alpha rays, Beta rays and the gamma rays and X rays come under the category of ionizing radiations while low penetration rays lik U.V. rays belong to non ionizing radiation.

Mayor (1920) experimented on the effects of X radiation *in Drosophila* (and similar other radiations) to study the effects on germplasm. He demonstrated that as a result of radiation, there was non disjunction of the X chromosome resulting in a XXY individual, which was female phenotypically.

H.J. Muller also worked on the effects of X rays on *Drosophila* and published a classical paper 'The artificial transmutation of the gene'. He also devised a technique to detect induced mutations (see later in the chapter, for detailed explanation).

Stadler studied the effects of X rays on barley plant, while Gager and Blackslee induced mutants in the jimson weed *Datura stramonium*. It should be noted however that X rays are lethal and the dosage of the rays given to the individual has to be carefully monitored. On an average it has been found out that 80% of the X ray induced mutations are lethal.

Oliver, Tirnofeeff Ressovsky and many others have worked out the relationship between the frequency of mutation and the amount of X radiation absorbed by the organism. The intensity of the radiation is measured in terms of Roentgen units or *r*- units. The frequency of mutation seems to be proportional to the amount of *r*- units absorbed by the plant. The duration of exposure or the distance between the radiation source and the experimental organism, however do not seem to affect the frequency of mutations. The following table gives the rates of mutation in *Drosophila melanogaster* due to two different doses of X rays. In the table below t4 dose is twice as large as the t2 dose. (The data in the table below are based on Muller's experiment, quoted from Principles of Genetics, Sinnott Dunn and Dobzansky).

Experiment	Number of chromosomes observed	Number of mutations		
		visible	semilethal	lethal
Control	198	0	0	0
Xrays t_2	676	1	44	9
Xrays t_4	772	3	12	89

Some of the mutations on exposure to X radiation in *Drosophila* are as follows.

Gene wt to W (white), W^e (eosin), and W^a (apricot). Genes coral (W^{co}) buff (W^b), cherry (W^c) and apricot W^a) can mutate into white (W).

Chlorophyll deficiency has been induced in barley due to X radiation (Stadler). X rays can also induce mutations in the wasp *Habrobracon*. Among the other types of ionizing radiation Alpha, Beta, and Gamma rays are important. The First two however do not seem to have a profound mutational role as they cannot penetrate the body cells as deeply as gamma rays.

Mechanism of action of ionizing radiation : There are two theories which account for the mechanism of mutagenic effect of ionizing radiation. These are the **direct hit theory** and the **indirect hit theory**. According to the direct hit theory, the radiation directly strikes the genes bringing about a change in the chemical composition. The electrons that. strike the atoms of genes bring about a change in the chemical composotion, thus altering the gene. But in several experiments, it has been seen that , the same dose of radiation does not result in the same frequency of mutation indicating that direct hit theory cannot account for all radiation induced mutations.

According to the indirect hit theory, gene alteration is brought about by changing the molecular environment of the surroundings. Muller is a supporter of this theory. According to him when the high speed protons of X ray collide with the molecules of the cell, electrons are ejected out of the atoms rendering them positively charged. The electrons that come out move at high speeds and knock out other electrons that come in their way resulting in an environment which is ionized (positively charged). The electrons moving freely may attach themselves to other atoms rendering them negatively charged. Various chemical reactions and shifts take place to bring both the positive and negatively charged atoms to their normal state. During this process, alteration in the chemical composition takes place, thus mutating the gene. At the molecular level, base pair sequences are altered in the DNA molecule (for more details see mutation at the molecular level, in the same chapter).

Effect of non ionizing radiation : Ultra violet rays are non ionizing because they have a long wave length and therefore have less energy. They are generally non penetrative. They may bring about many changes in the traits, but are not as effective as X rays. Altenberg was able to accelerate the rates of mutations by using U.V. rays. Stadler also achieved pollen mutations induced by U.V. rays in maize; the same is the case in *Antirrhinum* as reported by Noethling and Stubbs. The U.V. rays are known to cause alterations in the bond characters of purines and pyrimidines. Between the two, the latter are more prone to mutational changes.

Chemical mutagens : A number of chemicals are known to have mutagenic properties. It was C.Auerbach who discovered during second world war that chemicals could be used to induce mutations. Thorn and Steinberg (1939) had also reported the chemical induction of mutations. They treated the Fungus *Aspergillus* with nitrous acid. Since then a long list of mutangenic chemicals have been identified. Some of the common chemicals used as mutagens are given in the following table.

Table : Some mutagenic chemicals

Class of chemicals	Name of individual compound	Common Abbreviation
1 Aziridine	1 Ethyleneimine	EI
2 Mustards	2 Nitrogen mustard	
	3 Sulphur mustard	
3 Nitrosamines	4 Dimethylnitrosamine	DMN
	5 Nitrosoguanidine	NG
	6 Nitrosomethyl urea	NMU
4 Epoxides	7 Ethylene oxide	Eo
	8 Diepoxybutane	DEB
5 Alkyl sulphonates	9 Diethyl sulphate	dES
	10 Methyl methane sulphonate	MMS
	11 Ethyl methanal sulphonate	EMS
6 Others	12 Nitrous acid	
	13 Colchicine	
	14 Maleic hydrazide	
	15 Hydrazine	
	16 Hydroxylamine	
	17 Caffeine	
	18 Hydrogen peroxide	
	19 Formaldehyde	
	20 Phenol	
	21 Urithane	

Hydrogen peroxide is known to bring about mutations in *Neurospora.* Urithane and phenol are capable of inducing mutations in higher plants and *Drosophila.* Formaldehyde may induce mutations during spermatogenesis in *Drosophila.*

Colchicine is an alkaloid extracted from the bulbs of *Colchicum autmnale* of the family Liliaceae. This is known to induce polyploidy by suppressing spindle formation. Colchicine is not known to bring about any point mutation. Mustard gas is another chemical widely employed as a mutagen. Its mutagenic effect has been tested with reference to *Drosophila, Neurospora,* bacteria, barley, mice etc.

Temperature induced, mutations : Child, Plough and Ires have studied the effect of temperature as a mutagenic agent in *Drosophila.* A range of temperature from 4^{0}C to lethal range was used to find out the possible effect of temperature on the rate of mutation. A higher range of temperature is known to accelerate the rate of mutation. But it has to be borne in mind that a gradual increase in temperature might not be an effective mutagen. Any sudden increase or decrease in temperature might alter the gene copying mechanism thus altering the gene. The temperature effects largely remain as a somatic influence.

Vegetable oils as mutagenes : In one of their experiments concerning oil yielding plants, Swaminathan and Natarajan (1959) noticed that seeds with high oil content are generally resistant to mutagens. They soaked the cereal grains in oils of ground nut, mustard, and castor and wanted to know whether these oils provide any mutation resistance to the grains. Much to their surprise, the results were startling. Instead of the oils providing resistance to the cereals, they (oils) themselves had acted as mutagens as noticed by marked reduction in rates of germination, chromosome irregularities and many ear head variation mutations in wheat, one of the cereals used in the experiment. It was assessed that in wheat *(Triticum astivum)*, the rate of mutation, induced was much higher than that of X rays.

Detection of mutations : A variety of methods are available for detecting mutations, both in plants and animals. Some of these are described below.

CLB method for detecting mutations : This method perfected by Muller has been used to detect mutations in X chromosomes of *Drosophila.* Even though it is primarily meant to detect radiation induced mutations, it can be used in the detection of spontaneous mutations as well. Three characters on the X chromosome are chosen to study the mutation. These are -

C = Cross over suppressor. This represents a long inversion in the X chromosome, suppressing the cross over in the inverted portion. Cross over may still occur, but cross over gametes rarely survive due to deficiences and duplications.

L = Lethal gene. This is a recessive character. Hence homozygous females (l/l) and hemizygous (l/y) males do not survive.

B = Bar eye. This is a dominant gene. This is used as a phenotypic marker for the flies required to mate in the F_1 generation. The working mechanism of this is illustrated in the Fig. A CLB male is mated to a normal male which is subject to radiation. The idea is to find out whether there are any lethal mutations in the X chromosome of the male. Y chromosome is regarded as neutral as for as sex linked characters are concerned, hence the male will be hemizygous as for as genes on X chromosomes are concerned. Therefore any X chromosome of male which is irradiated can be followed in subsequent generations and lethal mutations if any can be detected.

In the X_1 generation (F_1 generation; called X_1 because the individuals are X radiated) the irradiated X chromosome of the male goes to the daughters, who have another X from the mother; hence in this generation any mutation induced can not be detected. The X_1 progeny will be as follows.

Females : 50% CLB; 50% Normal

Males : 50% CLB (die); 50% Normal.

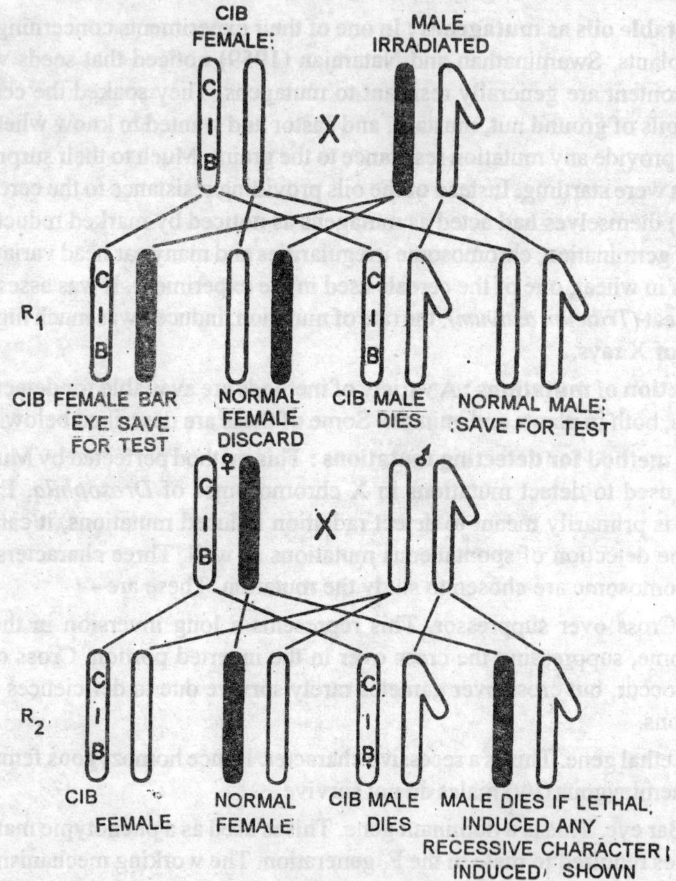

Fig. 10.2 Gene Mutation

Muller's CIB method to detect induced mutations drosophila melanogaster

A cross is now made between the normal male (with X also normal because it is obtained from female) and the CLB female. This CLB female has one of its X chromosome from the male parent which is irradiated and the other is CLB. In this (X_2 or F_2) generation all the males get one of the X chromosome, they die (50%) because there is already a lethal gene. If they get the irradiated X chromosome, and if a new lethal mutation is induced then the other 50% also will die. Hence if there is a lethal mutation on the X chromosome in the F_2, no male progeny will survive and all will be females. This is a clear cut evidence for the presence of a lethal mutation on the irradiated X chromosome. In the X_2 generation the progeny will be as follows.

Females : 50% CLB; 50% Normal (with an irradiated X chromosome)

Males : 50% CLB (die); 50% carry irradiated X chromosome; die if there is a lethal mutation.

The Muller 5 method : .Like the CLB method this is also devised by Muller. The experiment involves crossing a homozygous Muller 5 female *Drosophila* with an irradiated male. The Muller 5 female *Drosophila* has two marker genes - apricot eye (w^a/w^a) and Bar eye (B/B) in a long inversion so that there is no cross over. A homozygous apricot (eyed) Bar eyed female is crossed with an irradiated male. As shown in the figure, the F_1 generation female offspring which gets the irradiated X from the male parent is crossed with a wild male. If a lethal mutation has occurred in the X chromosome, no wild types of male would survive, because they would get the irradiated X.

The absence of wild males in the F_2 is an indication of lethal mutation. Muller was awared Noble prize in Medicine in 1946 for this and many other discoveries that he had made in Genetics.

Detection ofAutosomal mutations (Balanced lethals)

Autosomal mutations can be detected by making use of balanced lethal genes in organisms. A balanced lethal stock is one which carries a lethal gene in one of the homologues and its wild allele on the other homologue. Homozygous individuals will die, so what survives is always heterozygous. Balanced lethals can be studied by using a specific strain of *Drosophila melanogaster* called the **curly lobe plum.**

In this, a stock of flies carry on one of the autosomes two dominant genes *C* (curly wing) and *L* (lobed eye) and a recessive lethal gene. An individual with two *CYL* chromosomes will not survive because the recessive lethals will be in double dose and its homologue carried the *Pm* (plum coloured eye) gene which is also dominant.

When two individuals with the genotype *{(CYL/Pm)* X *(CyllpM)}* are crossed, the heterozygous individuals only would survive, as homozygous for *(CYL/CYL)* would carry lethal genes in double dose, because *CYL* chromosome carries a lethal gene. There is a cross over suppressor to prevent the transfer of the gene *CY to* the chromsome carrying *Pm* gene.

A female *CYL/Pm* is crossed to another normal male which is irradiated.

Four types of progeny are produced in the F_1. These are -

(a) Curly lobe (with an irradiated autosome A)

(b) Curly lobe (with another irradiated autosome B)
(c) Plum (with an irradiated autosome A)
(d) Plum (with another irradiated autosome B)

One of these flies is selected (because each one has one irradiated chromsome) and back crossed to curly lobe plum.

The progenies will be of the following types.

(a) homozygous curly lobe (dies)
(b) Curly lobe plum.
(c) Curly lobe with an irradiated chromsome.
(d) Plum with an irradiated chromsome.

Among the above progeny, (c) flies are crossed among themselves to obtain -

(a) homozygous curly lobe (dies)
(b) Heterzygous curly lobe (with an irradiated chromosome)
(c) same as b.
(d) A fly with both irradiated chromosomes.

It is in the (d) category that any mutation induced would appear. If the induced mutation is lethal, it also dies and only the curly lobed individuals survive. If the mutation induced is not lethal, but a recessive trait, that would be expressed, since it is homozygous. The same test can be done by selecting the plum with an irradiated chromosome and crossing it with a curly lobe plum with an irradiated chromosome.

The significant aspect of this cross is to bring together the same irradiated chromosome in the homozygous condition to detect all the recessive mutations.

Detection of induced mutations in plants

L.J. Stadler of the University of Missouri and Ralph Singleton of the University of Virginia have worked on the methods to detect mutations in maize and barley.

Stadler (1942) has made a detailed study of spontaneous mutations in maize. The following are the important steps.

(a) A genetic stock dominant for a number of genes is grown as the female parent. The tassels are removed before pollen discharge (to avoid self pollination).
(b) A multiple recessive plant in every fifth row is used as the pollen providing parent for (a) plants.
(c) Seeds were observed in large numbers for endosperm characters. Most of them show dominant characters. If any recessive character (of the male parent) appeared it is due to mutation in the female gametes.

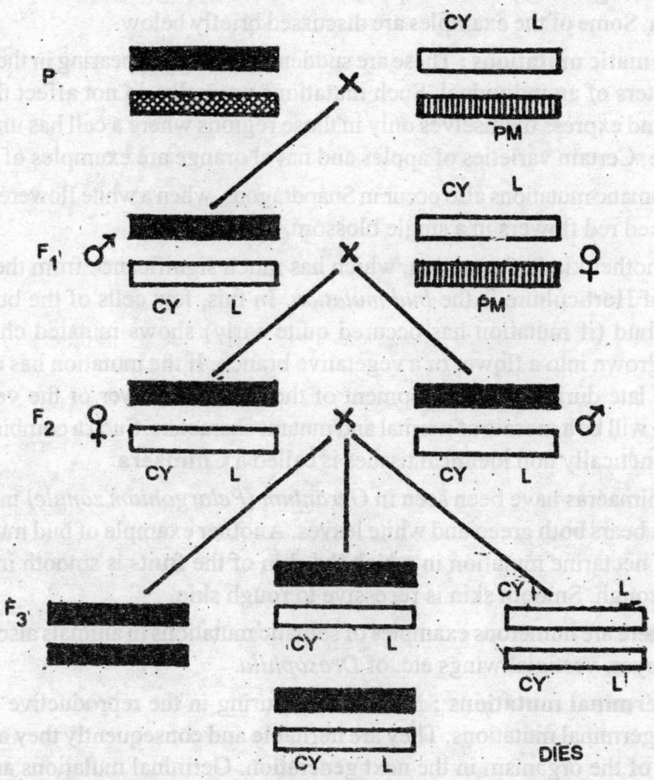

Fig. 10.3 Gene Mutation
Detection of lethal mutations in a balanced lethal system

Singleton (1962) used a techinque similar to that of Stadler do detect both spontaneous and induced mutations in the pollen of maize. His method involves collecting pollen from plants dominant for many genes and cultivated in a field irradiated with cobalt 60 and transferring them (pollen) on to the silks of plants with all recessive traits and cultivated in a normal field. The seeds of the progeny would normally show dominant characters. If any mutation has been induced in a dominant gene, the recessive trait would appear in the progeny. In this way Singleton studied four genes *Su -su,* on chromosome 4, *Pr - pr* on chromosome 5, *Y-y* on chromsome 6 and *Sh -sh* on chromosome 9. The phenotypes of these genes are *Su* - normal endosperm, *su* - sugary endosperm; .Pr- Purple aleurone,pr-red aleurone; *Y*- yellow endosperm, *y* - white endosperm and *Sh* - full endosperm, *sh* - shrunken endosperm.

Apart from spontaneous and induced, mutations may also be classifed (see the beginning of this chapter) into various categores based on different criteria. Some of the examples are discussed briefly below.

Somatic mutations : These are sudden alterations appearing in the somatic characters of an individual. Such mutations normally do not affect the entire plant and express themselves only in those regions where a cell has undergone change. Certain varieties of apples and navel orange are examples of this.

Somatic mutations also occur in Snapdragons, when a white flowered variety produced red flowers in a single blossom.

Another kind of mutation, which has much significance from the poin of view of Horticulture is the *bud mutation.* In this, few cells of the bud or the entire bud (if mutation has occured quite early) shows mutated characters, when grown into a flower or a vegetative branch. If the mutation has occurred a little late during the development of the bud, the flower or the vegetative branch will be a mosaic of normal and mutant characters. Such a combination of two genetically non identical tissues is called a **Chimaera.**

Chimaeras have been seen in *Geranium (Pelargonium zonale)* in which a branch bears both green and white leaves. Another example of bud mutation is that of nectarine mutation in which the skin of the fruits is smooth instead of being rough. Smooth skin is recessive to rough skin.

There are numerous examples of somatic mutations in animals also; for e.g. white eyes, vestigial wings etc. of *Drosophila.*

Germinal mutations : Mutations occuring in the reproductive cells are called germinal mutations. They are heritable and consequently they affect the whole of the organism in the next generation. Germinal mutations are called gametic when they occur in gametes and zygotic when they occur in the zygote. The gametic mutation will express itself readily when either the mutation is dominant or if recessive two mutant gametes should unite. Zygote mutations on the other hand express themselves readily. The ancon breed sheep is a typical example of germinal mutation.

Biochemical mutations : Mutations bringing about alterations in metabolic pathways, nutritional requirements etc. are called biochemical mutations. These are mostly noticed in microorganisms. Biochemical mutations in *Escherichia coli* can be induced by repeated sub culturing. In *E.coli,* the normal strains are called *Paratrophs* while the mutants are called *Auxotrophs.* These mutations are for some of the nutritional requirements like Biotene, Methionine, Leucine etc. The paratroph is called genetically B+, M+ and L^+ because it has the capacity to synthesize all these. On mutation the auxotroph loses the capacity to synthesize these and consequently it is known as $B^- M^- L^-$

Several metabolic disorders due to biochemical mutations have been reported in man also. They are commonly called **Inborn errors in metabolism'**. The first such biochemical disorder to be reported was Alkaptoneuria (Garrod 1902). This relates to a metabolic disorder concerning the amino acid phenyl alanine. **Alkaptaneuric** individuals have a mutated gene which does not produce the .enzyme homogentisate oxidase, resulting in the accumulation of homogentisic acid in the cell, which is excreted along with the urine. Homogenticic acid is a strong reducing agent and gets oxidised in the presence of air into a black pigment, resulting in the black colouration of the urine.

Another biochemical mutation in man is phenyle ketoneuria. It is caused by a recessive mutant gene in which due to failure of conversion of phenylalanine to tyrosine, the former accumulates in the blood and excreted through the urine. This leads to mental retardation. In addition to the above, Tyrosinosis, Albinism and Cretinism are the other metabolic disorders due to biochemical mutations.

Reverse mutations : The back mutation of a gene to come back to its wild state is known as reverse mutation. In *Drosophila,* the wings are normally long and due to a mutation, the wing becomes miniature. In some instances, all of a sudden, a few long winged flies appear in a population of miniature winged flies. This is due to the reverse mutation of the miniature winged gene back to the wild state. Another reverse mutation in *Drosophila* is the mutant, fork bristles back to normal bristles. There are many such reverse mutations in bacteria and fungi.

Lethal mutations : Mutations of gene, from normal to lethal leads to the death of the individual. Normally lethal genes are recessive and nothing happens to the organism in the heterozygous condition. But occassionally there may be dominant lethals in which case individuals do not survive even in the heterozygous condition. The latinum colour in Fox, yellow coat colour (see chapter on Interaction of factors for a detailed discussion) in mice are such examples.

Many lethal mutations are also known in plants *likeAntirrhinum majus* (Snapdragon). The golden plant stock has variegated leaves. On self pollination this variety produces Golden and green in the ratio 2:1 (not 3:1), obviously the golden is heterozygous and green breeds true. The other one (golden) is missing because pure golden can not exist due to homozygosity of lethal genes. Lethal genes are also known in paddy and maize plants.

MUTATIONS AT THE MOLECULAR LEVEL

The basic hereditary material namely DNA is a highly stable biological molecule. The hereditary information is present in the form of sequence of the four nucleotides Adenine, Thymine, Guanine and Cytosine. During the inheritance of characters from one individual to the other, the sequence is duplicated and transmitted in a precise and accurate manner.

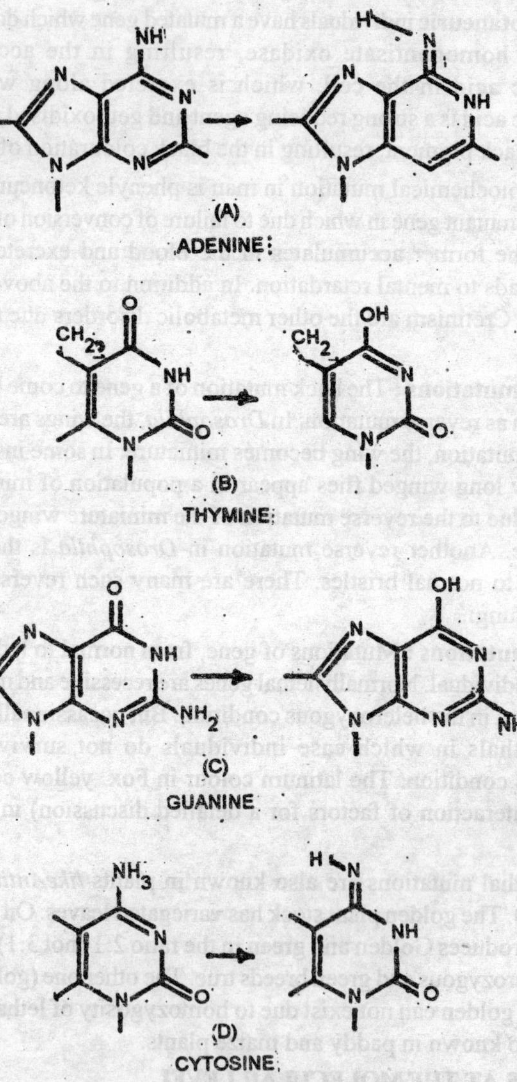

(A)
ADENINE

(B)
THYMINE

(C)
GUANINE

(D)
CYTOSINE

Fig. 10.4 Gene Mutation
Uncommon forms of DNA bases

A slightest mistake at this stage will alter the base pair sequence and results in an entirely different type of genetic code resulting in the formation of an altered protein which may have far reaching effects on the organism. Such mistakes or faults occuring in the DNA molecule during duplication are ultimately the basis of mutation at the molecular level. The following two basic types of sub nucleotide changes are noticed in the DNA molecule - Frame shift mutation and substitution mutation.

Frame shift mutations : Addition or deletions of nitrogen bases either from DNA or mRNA are called frame shifts, because, from the site of the alteration, the reading frame of the message alters. In some cases frame shifts are neutralized ie. deletion may be neutralized by the addition at the same site. Frame shift mutations are of two categories. These are deletion mutations and insertion mutations. In deletion a single nucleotide in a triplet in deleted. These are found frequently in T_4 bacterophages. In insertion mutations an extra nucleotide is added to a gene. These insertions can be induced by chemicals such as acirdine, Proflavin etc.

Substitution mutations : This generally occurs during the replication of DNA, when one nitrogenous base is replaced by another changing the triplet codon. The altered codon now codes for a different amino acid resulting in an altered protein. There are two kinds of substitutions - Transitions and Transversions.

Transitions : The replacement of one purine by another purine or one pyrimidine by another pyrimidine is called trasition. In other words A is replaced by G and T by C or *vice versa.*

Transitions may be brought about by tautomerization, ionization, base analogs or by deamination. In tautomerization, a normal base, pairs with its tautomer partner [A tautomer is an alternate state of existence in a molecule. For e.g. In cytosine and adenine tautomers, the normal amino group NH_2 is converted into the NH (imino) group and the keto group (C = 0) of thymine and Guanine is converted into the OH group in their tautomers. Normally this does not happen]. Such pairing is called *forbidden base pair.* Watson and Crick (1953) opined that the existence of a tautomer to DNA provides a mechanism for mutation.

Ionization is another mechanism by which transition can be brought about. Ionization of a base at the time of DNA replication involves the loss of H from a nitrogen base. Such ionized bases pair with normal bases.

Another method by which transition may be brought about is by the replacement of normal bases by their analogs. These compounds called ***base analogs*** are derivatives of the natural bases. For eg. cytosine may be replaced by ***5 hydroxyl methyl cytosine*** or by ***5 glucosyl hydroxy methyl cytosine.***

There are naturally occuring base analogs of cytosine found in T even phages of *E.coli. Uracil* has a naturally occuring analog - 5 *hydroxy methly uracil* found in some Viruses. Besides the naturally occuring analogs, artificial analogs like 5 *bromouracil, 5 iodouracil, 5 bromocytosine, 5 methyl cytosine* may be used as analogs of thymine (first two) and cytosine (last two) which may be used to induce mutations.

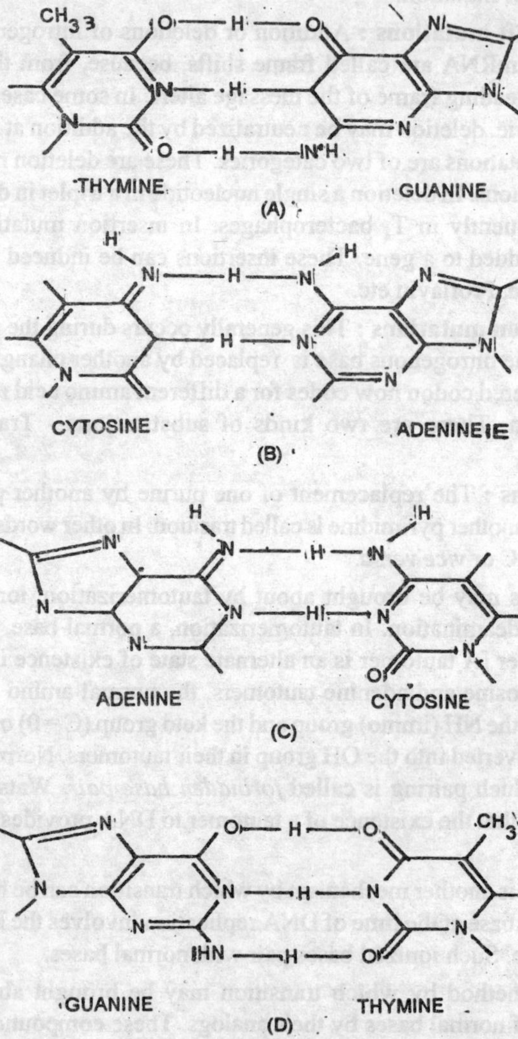

THYMINE GUANINE

(A)

CYTOSINE ADENINE

(B)

ADENINE CYTOSINE

(C)

GUANINE THYMINE

(D)

Fig. 10.5 Gene Mutation

Unusual pairing of tautomeic bases in DNA

Deamination is another method of induction of transition, by replacing the NH_2 group of bases with OH group. This may be induced by chemials such as nitrous acid, hydroxylamine, diethyl sulphate etc. Deamination of cytosine leads to the formation of uracil, and that of adenine forms hypoxanthine. During replication U pairs with A and hypoxanthine pairs with C. As a result A = T is substituted by G = C and G=C byA=T.

Trasversions : Unlike m transition, a purine is replaced by a pyrimidine and *vice versa* ie. CG replaced by GC and AT by TA. This change can be induced by alkylating agents like ethylmethane sulphate (EMS) and Methylmethane sulphate (MMS). The alkylating agents alkylate the nitrogen in the 7th position in a purine and lead to its separation from the DNA strand. This gap can be filled by any of the four bases at the time of replication. Trasversions were first postulated by E.Freese in 1959.

PRACTICAL APPLICATION OF MUTATIONS

Generally mutations are lethal and recessive and possibly of no help except as a scientific curiosity in understanding the basic funcitoning of the hereditary material. But there are specific instances where both spontaneous and induced mutations are of selective value from the point of view of Evolution and also from the point of view of human beings. Ancon breed sheep is one such example of a spontaneous mutation. Mutations are also of value in plant breeding, when mutant varieties can be stabilized in the gene pool by cross breeding. Gustaffson has estimated that about one in thousand mutations may be used in breeding experiments. Several mutations in wheat, like branched ears, resistance to lodging, high lysine content, amber seed colour, awned spikelets have been useful inbreeding. The amber seed colour which is originally seen in the mexican variety has been induced in wheat of Indian varieties.

Many useful mutations in rice are also known. Some of these are *Reirnei* variety mutant, for reduced duration of the crop. In barley also mutants like *erectoids* and *eceriferum* have been induced for increased yield. The *Todds Mitchan* variety of mint plant *and-Aruna* variety of castor serve as good examples of the role of mutations in crop improvement. The above mentioned variety of mint has disease resistance as well as high oil content. In the castor variety duration of the crop was reduced by 130 days (originally 270 days). In a survey conducted by FAO in 1974 it has been estimated that 98 varieties of crops have been improved by induced mutations.

11

GENETIC ENGINEERING

Recombinant DNA technology or Genetic engineering as it is popularly known, is the newest tool that is available in the human arsenal of knowledge, in man's quest for a better human life. What is genetic engineering or recombinant DNA technology?

Imagine a situation of chronic diabetic patients being able to get back the capacity of synthesis of insulin; patients suffering from incurable diseases getting an injection of fene therapy–lo and behold they are cured of their diseases; consider a situation which is at present in the realms of fantasy of making man producing his own food just like the green plants do from natural sources; or visualize a situation wherein revolution in agricultural technology would make the use of chemical fertilizers completely redundant or going to the extreme, culturing a human being of the desired trait on a medium of nutrients i.e., man made to order. The scope is fantastic; possibilities are immense; it is as if the role of creator has been for once given to human beings. Well, this is genetic engineering or recombinant DNA technology, the most powerful biological technique which the human beings can use either to make or mar the society either for progress or for destruction. Using the techniques of genetic engineering, you can create deadly pathogenic bacteria and unleash them on human beings; you can produce varieties of agricultural crops which are disease free or you can introduce pathogens into the crops and totally destroy them. Genetic engineering used judiciously can be a harbinger of new revolution but used injudiciously may produce many prototype of criminals, antisocial elements, boot leggers etc. In this chapter, we shall briefly discuss some of the fundamental aspects of genetic engineering and how to use them for betterment of human life.

Genetic engineering may be defined as the method of artificial synthesis of new genes and their subsequent transplantation in the new genome of an organism, by molecular biological techniques. Recombinant DNA is achieved by constructing molecules outside the living cells by joining natural or synthetic DNA molecules that can replicate inside a living cell. Recombinant DNA technology involves several stages or steps.

The main aspects of genetic engineering (gene cloning) are the following –

1. Identification and isolation of the gene to be cloned.
2. Insertion on the gene into another piece of DNA called vector which will allow it to be taken up by microorganisms like bacteria and replicate within them as the cells divide.
3. Transfer of the recombinant vectors into bacterial cells.
4. Selection of those cells which have the desired recombinant vectors.
5. Growth of the microorganisms to get increased quantity of cloned DNA.
6. Expression of the desired gene to obtain the desired product.

Outline procedure of genetic engineering

Genetic engineering of a desired gene and gene product involves the following procedure. First of all, the desired gene has to be obtained from a source which then has to be identified and isolated.

(i) The living cells from which the genetic material is to be obtained has to be broken open. One popular method is to shear the cells in a blender and treat them with a suitable detergent.

(ii) The genetic material from the cell is to be removed. Several techniques are available to remove the DNA. The DNA molecules can even be spooled on to a glass rod. The glass rod bearing the DNA molecules is then lifted up from the debris of the cells.

(iii) The next step is to cut away the specific genes of interest from the DNA molecule. Restriction enzymes popularly called molecular scissors are used to cut and separate the desired gene from the DNA helix.

(iv) In the next step, the specific sections of DNA are incorporated into cloning vectors or vehicles such as plasmids, phages etc. These cloning vehicles are then introduced into a rapidly multiplying living system such as bacteria where multiple copies of the cloned DNA can be obtained. The cloned DNA also called chimeric DNA, contains the desired foreign gene. The coloned DNA is also called recombinant DNA.

(v) In the final step which is a very delicate process, the cloned gene is allowed to express itself in the bacterial cell. Gene expression here means production of a desired gene product. For instance, if a human insulin gene is cloned in a plasmid and is inserted into a bacterial cell, the bacterial cell along with its metabolic products also will produce insulin. This insulin may then be separated by the normal methods of chemical purification and isolation.

RESTRICTION ENZYMES (restriction endonucleases)

The first account of restriction enzymes was given by Werner Arber while studying the bacteriophages. He found that in some instances as soon as the viral DNA entered into a bacterial cell, it (viral DNA) was cut into bits and destroyed. Based on this observation, Arber opined that the bacterial cell must possess certain specific DNA destroying enzymes that could infect only the viral DNA but not the host DNA. In otherwords, the enzyme should be such that it should be able to distinguish between foreign DNA and its own DNA. He gave the name restriction enzymes and hypothesized that these restriction enzymes can recognise and act at specific sites on the viral DNA.

The next step in the discovery of restriction enzymes was taken by Smith *et al* (1970), when they isolated a restriction enzyme from the bacterium *Haemophilus inflenzae.* Smith showed that the restriction enzyme can cut the viral DNA at specific sites. All the DNA termini were pGpTpPy-3' and 5' pPupApCp, showing that these were cut from the double stranded DNA having the following sequence

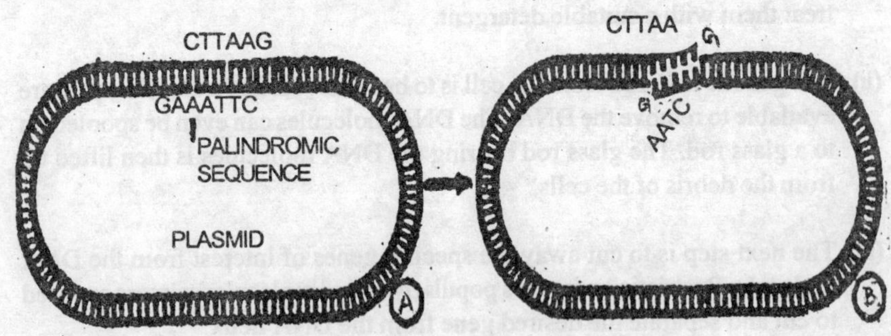

Fig. 11.1 Genetic Engineering
Action of restriction enzymes

5'pGpTpy	|	pPupApCp-3'
3'pGpApPu	|	pPypTpGp-5'

The above example shows that the restriction enzyme from *Haemophilus influenzae* can cut the DNA molecule which exhibit a two fold symmetry i.e., with a palindromic sequence (same sequence running in opposite directions),

some restriction endonucleases, their source, recognition sequence and sites of cleavage (indicated by arrow).

Enzyme	Source	Recognition sequence and clearing site	
		↓	
Eco R1	*Escherichia coli* Ry13	G.AA	TTC
		GTT	AAG
			↑
Hind II	*Haemophilus influenzae*	GTPy ↓	PuAC
		CAPu ↑	PyTG
Hind III	*Haemophilus influenzae* Rd	A AG	CTT
		TTC ↓	GA A
			↑
Hpa I	*Haemophilus parainfluenzae*	GTT ↑	AAC
		CAA ↓	TTG
Hpa II	*Haemophilus parainfluenzae*	CC	GG
		GG	CC
		↓ ↑	
Bam HI	*Bacillus amyloliquefaciens*	G GA	TCC
		CCT	AGG
		↓	↑
Bgl III	*Bacillus globigi*	A GA	TCT
		TCT	AG A
			↑
Hae II	*Haemophilus aegyptius*	Pu GG	GCPy
		Py CG ↑	CGPu
Hae III	*Haemophilus aegyptius*	GG ↓	CC
		CC ↑	GG
			↓
Hha I	*Haemophilus haemolyticus*	GC	GC
		C G	CG
		↑	
Pst I	*Providencia stuartii*	CTG	CA G
			↓
		GAC	GTC
		↑ ↓	

Enzyme	Source	Recognition sequence and cleaving site
Sma I	*Serratia marcescens*	CCC GGG GGG CCC ↓ ↑
Taq I	*Thermus aquaticus*	T C GA AG CT ↑
Bal I	*Brevibacterium albidum*	TCG ↓ CCA ACC ↑ GGT ↓
Sal I	*Streptomyces albus* G	G TC GAC CAG CTG ↓↑
Xor II	*Xanthomonas oryzae*	CGA TCG G CT AGC ↑
Alu I	*Arthrobacter luteus*	A G CT TC GA ↓
Msp I	*Moraxella* sp.	CC GG GG ↑ CC ↓
Mbo I	*Moraxella bovis*	GA TC
Sau 3 Al	*Staphylococcus aureus* 3A	GA ↓ TC CT ↑ AG

Subsequent work by Daniel Nathans (1971), showed that the restriction enzyme discovered by Smith can cut the monkey virus DNA into eleven well defined pieces. By using specific restriction enzymes it is also possible to map the DNA which they cut as their specificity is known. Arber, Nathans and Smith were awarded the 1978 Noble prize in physiology and medicine for their work on restriction enzymes.

In many microorganisms, restriction enzymes are a means of protection against intrudent viral DNA. In the working of the restriction enzyme, the nucleotide sequence of the host DNA is identified and methylated to avoid degradation. As a result of this the DNA gets the protection while the foreign DNA with the same sequence will not be methylated and coonsequently subjects itself for degradation. The whole mechanism is called the RM system (restriction modification), and is wide spread in bacteria.

Classification of restriction enzymes

Restriction enzymes are able to recognise specific nucleotide sequence in the double stranded DNA and cleave both the strands. To date, more than 150 restriction enzymes have been discovered which are known to recognise 40 different nucleotide sequence. Some restriction enzymes, however, may not cut DNA at specific sites. Based on this, restriction enzymes are classified into two, categories, type 1 and type 2. Type 1 restriction enzymes recognise a specific sequence of DNA molecule but the cut is made elsewhere whereas type 2 enzymes make a cut only within the restriction site and produce two single strand breaks, one break in each strand. Type 2 enzymes are most important because they produce broken chain of nucleotides with a palindromic sequence. The following table gives a summary of the difference between type 1 and type 2 eyzymes.

	Type 1	Type 2
1.	Non-specific in cleavage	Cleavage is site-specific
2.	M.W 300,000	Lower MW (20,000 - 1,00,000)
3.	Has non-identical subunits	Has two identical subunits
4.	Require ATP, Mg ++ and generally S-adenosyl	Require only Mg ++ as co-factor
5.	Examples : EcoB, EcoK, EcoPi	Example : EcoRI

Nomenclature of restriction enzymes

The names are designated according to the following principles.

(1) Name of the organism is identified by the first letter of the gene's name and the first two letters of the species name to form a three letter abbreviation in italics. For example *E.coli* = *Eco* and *H.influenzae* = *Hin* etc.

(2) A strain or type identified is written as a subscript. For example, *E.coli* strain k is written as Eco_k.

(3) In cases where the restriction modification systems are genetically specified by a virus or plasmids the extra chromosomal element is identified by a subscript as *Eco RI, Eco PI* etc.

When a particular strain has several restrictions and modification systems, these are identified by Roman numerals. For example Hin dI, Hin dII, Hin dIII etc.

Cleavage brought about by restriction enzymes

Restriction enzymes have an ability to recognise a particular sequence and cleave DNA molecule at that site. Most of the sequences recognised by these enzymes are tetranucleotide or hexanucleotide pairs which show a characteristic two fold symmetry i.e., palindromic sequence. The Eco RII site is very unique as it contains 5 base pairs with the axis of symmetry passing through the middle AT pair as shown below.

CCTGG
CCACC

Some times restriction enzymes from unrelated species can recognise the same nucleotide sequence. For instance, GGCC sequence specific enzymes have been discovered in *Bacillus subtilis, Haemophilus aegypticus* etc. Such enzymes are called isochizomers.

Cleavage sites may be even or staggered. In even cleavage the breaks on the two strands are exactly opposite to each other. Whereas, in staggered breaks, the breaks on the two strands are a little away from each other. Even breaks do not produce unpaired terminal nucleotides, whereas staggered breaks may produce sticky ends.

Use of restriction enzymes

The use of restriction enzymes has proved to be a highly useful tool for the sequence analysis of DNA. It also plays a significant role for gene cloning and amplification in genetic engineering.

The DNA molecule from bacteriophages and many other viruses consists of 5,000 to 50,000 base pairs. It is of fundamental importance to know the sequence of these bases to understand the gene structure and regulation and also for decoding the information stored in these genes. By the use of several restriction enzymes it is possible to cut fragments of DNA as small as 300 bp. These fragments can be used for sequence analysis and mapping.

The DNA segments produced by restriction enzymes can be covalently linked to the plasmid DNA and such recombined DNA can be inserted into bacteria such as *E.coli* for transformation. The transformed cells can now be cultured as separate colonies.

The ability of restriction enzymes to locate and cut at a particular sequence is of great help in identifying and isolating a particular gene.

Identification of a required DNA piece (gene)

Theoretically, identification of a gene involves, the reading of the message back in protein synthesis. During protein synthesis DNA produces mRNA and this in turn identified and allows the alignment of amino acids to synthesize a protein. In the identification of a gene responsible for the production of a particular protein, the process has to be followed backwards. Let us take an instance of the human insulin which is a protein.

The steps involved in the identification of the gene or segment of DNA responsible for insuline are as follows.

(a) Isolation and purification of the protein.

(b) Sequential degradation of the protein : In sequential degradation, not only the qualitative composition of the amino acids are known, but even their sequence of arrangement in the polypeptide chain will be known. After ascertaining the sequence of amino acids, the next step would be :

(c) Identification of codes for each amino acid : This can be done easily as we know the triplet codes for each amino acid. This would give us a length of the mRNA segment equivalent to the length of the polypeptide chain. (For example, if the protein consists of 25 amino acids, the length of the mRNA should be 75 nucleotides).

The nucleotide sequence in mRNA can be easily substituted for the complementary DNA segment. We now have a gene or a segment of DNA which governs the production of insulin.

ISOLATION OF SPECIFIC DNA SEGMENT (GENE)

The DNA of an organism, whose specific gene is required for the purpose of expression in another organism contains thousands of genes. The first problem therefore, is to separate the required gene from the DNA of the organism. There are several techniques to isolate the required gene from the DNA molecule.

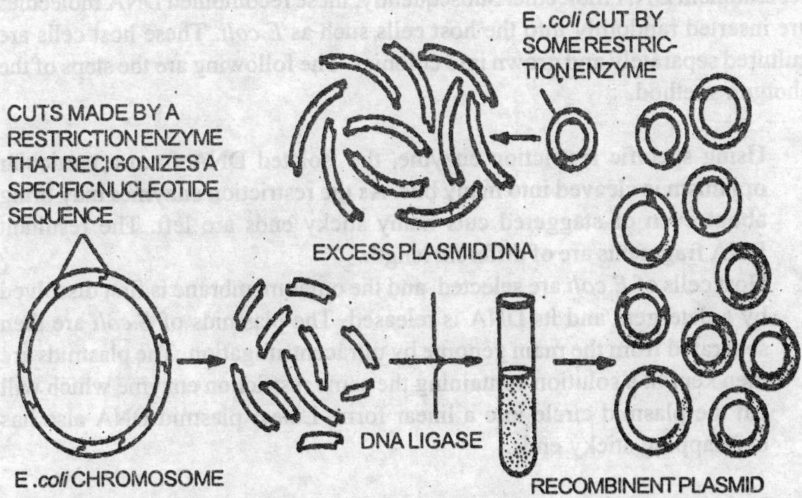

Fig. 11.2 Genetic Engineering
Isolation and recombination of genes using shot gun method

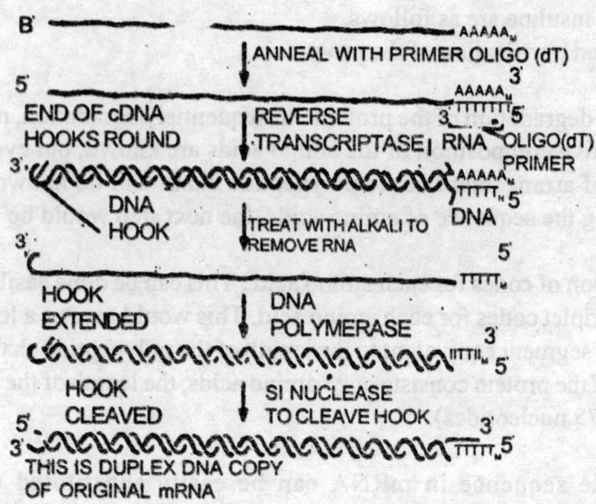

Fig. 11.3 Genetic Engineering
Obtaining cDNA from mRNA

Shotgun method

In this, the entire DNA of the organism is subjected to the action of restriction enzymes and is broken down into a number of fragments. Each of these fragments is then combined with the DNA from a suitable vector to produce recombinant DNA molecule. Subsequently, these recombined DNA molecules are inserted randomly into the host cells such as *E.coli*. These host cells are cultured separately and grown into colonies. The following are the steps of the shotgun method.

1. Using specific restriction enzyme, the isolated DNA from a particular organism is cleaved into many bits. As the restriction enzymes may bring about even or staggered cuts many sticky ends are left. The resultant DNA fragments are of different length.
2. Host cells of *E.coli* are selected, and the outer membrane is first dissolved by a detergent and its DNA is released. The plasmids of *E.coli* are then separated from the main genome by ultracentrifugation. The plasmids are then kept in a solution containing the same restriction enzyme which will cut the plasmid circle into a linear form. Linear plasmid DNA also has overlapping sticky ends.
3. In this step, isolated DNA and linear plasmid DNA are mixed. As a result, random association occurs between the foreign DNA and the plasmid DNA. Subsequently, the foreign DNA gets integrated into the plasmid

DNA to form a long linear DNA molecule, which then becomes a circle due to covalent joining. The resultant DNA is a recombinant or chimeric DNA.

4. The chimeric DNA molecules are now placed in a solution of calcium chloride which also has *E.coli* cells. On heating suddenly upto 42°C, the membranes of bacteria become permeable to chimeric DNA molecules. Thus, the *E.coli* cells incorporate the chimeric DNA plasmids.

5. The bacterial cells containing in the chimeric plasmids is then cultured on a medium. The cells express not only their genes but the genes of the chimeric plasmids also, thus producing a desired product. All these cells are produced by the division of the original bacterial cell having a single chimeric plasmid. Hence, they are called clones. Insertng a chimeric plasmid into a bacterial cell is of a very rare occurrence; it has been estimated that only one is 10^5 of 10^6 cells are capable of undergoing transformation.

In shotgun method, several genes can be cloned at a time and it is an omnibus method because the genes to be cloned are not picked and identified before.

Obtaining DNA from mRNA

In shotgun method, gene cloning is a general one and not specific as the genes to be cloned are not identified and isolated before. Hence, the chances of inserting a particular gene by this method are very rare. Anothei specific method of isolating gene is to first obtain the messenger RNA transcribed by that particular gene and then obtain DNA from mRNA. DNA from mRNA can be obtained by the following two methods.

Hybridization method

When DNA transcribes and forms mRNA, a complex is formed. The hybridization method depends on this mechanism. At the first step, the total DNA from an organism is isolated and treated with heat or alkali to convert it into a single stranded form. These single strands are then mixed with the specific mRNA obtained from the gene. Obviously, the mRNA chains pair with that section of DNA which has complimentary nucleotides to it, thus forming aDNA-RNA hybrid. This complex can be isolated and the DNA can be separated from the RNA. The single strand DNA can be converted into a double strand DNA by enzymes such as polymerase I. This method is highly useful for isolating ribosomal genes.

Reverse transcriptase method

In the normal transcription, the flow of information is from DNA → RNA → proteins. Transcription of RNA from DNA is catalysed by the enzyme transcriptase also known as DNA dependent RNA polymerase. Temin and Baltimore independently discovered that occasionally the genetic information may flow in the reverse direction also (from RNA to DNA), in some RNA tumour viruses. They named the enzyme *Reverse transcriptase* (RNA directed DNA polymerase). This enzyme catalyses the synthesis of a single strand DNA using an RNA template. This discovery has proved to be highly useful for isolating particular genes.

In reverse transcriptase method, purified mRNA segments of the required gene are used as a template to produce a fragment of single stranded DNA under the influence of the enzyme reverse transcriptase. As this single strand DNA is complementary to mRNA, it is called complimentary DNA or cDNA. Hydrolysis of RNA template will separate a single strand DNA. This single strand DNA is converted into double strand using the enzyme DNA polymerase. Another enzyme S1 nuclease breaks the covalent linkage between two DNA strands.

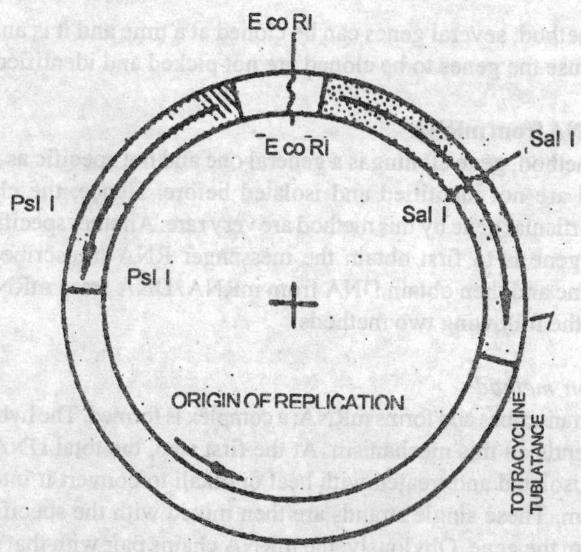

Fig. 11.4 Genetic Engineering
Genetic map of plasmid pBR 322

VECTORS (Carriers or Vehicles)

Vectors are defined as those DNA molecules that can carry a foreign DNA fragment inserted into them without any disturbance in their expression. Vectors are also called vehicle DNAs. Vectors are obtained from a number of sources. These are grouped into categories such as plasmids, bacteriophages, cosmids and phasmids.

Plasmids

These are extra chromosomal, self- replicating and double stranded covalentaly closed circular DNA molecules present in the bacterial cell. A number of plasmids are present in bacterial cells that have sufficient genetic information for their replication. Eventhough they are not necessary for the survival of the bacteria, they carry such necessary information as drug resistance, metal resistance, toxin production, nitrogen fixation etc. Naturally occuring plasmids can also be modified by *in vitro* techniques such as code transformation. Use of plasmids as a vector for DNA cloning was reported for the first time by Cohen et al (1973).

Plasmids are used as cloning vehicles when they possess the following features.

1. It can be readily isolated from the cells.
2. It possesses a single restriction site for one or more restriction enzyme(s).
3. Insertion of a linear molecule at one of these sites does not alter its replication properties.
4. It can be reintroduced into a bacterial cell and cells carrying the plasmid with or without the insert can be selected or identified (Bernard and Helinski, 1980).
5. They do not occur free in nature but are found in bacterial cells.

In many instances, the main objective of coloning is the insertion of a particular fragment of DNA into a suitable plasmid vector and its amplication. Amplication is a process of increasing the number of plasmids in a bacterial cell. In this process, a cell containing a relaxed plasmid is treated with a drug to inhibit protein synthesis. Consequently, replication stops and the relaxed plasmid (eg pBR 322), continues to replicate despite the drug treatment. Addition of chloramphenicol causes the plasmid BR 322 to increase its number to 3,000 per cell. In order to replicate and clone a piece of foreign DNA, it is necessary to possess a sequence of nucleotides which is recognised by the host bacterium as origin of replication. The origin of replication occurs naturally in plasmids. In naturally occurring plasmids, due to exchange of genetic information between plasmids and bacterial chromosome new genes originate in the plasmids. Therefore, plasmids have been artificially developed with specific sequence of genes so that they would integrate foreign DNA without any problem.

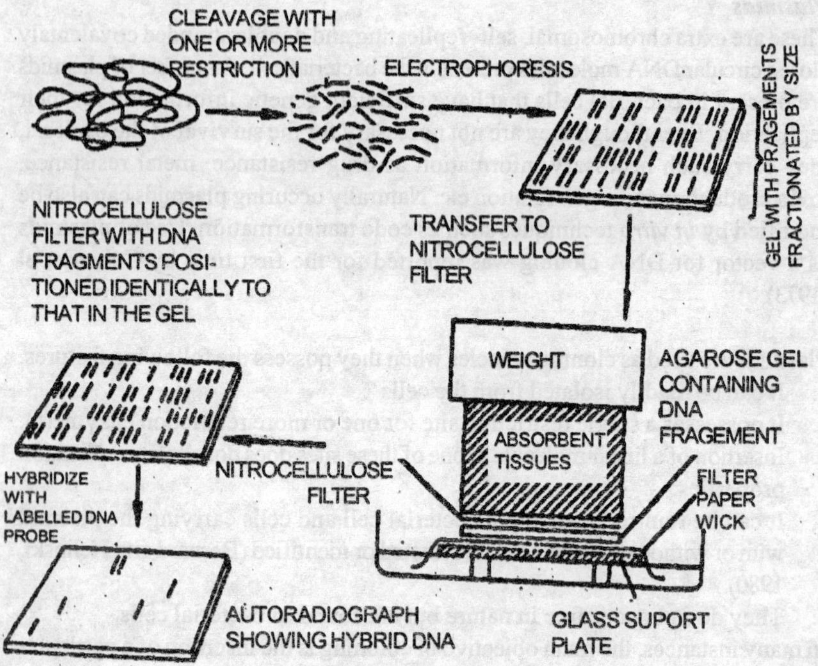

Fig. 11.5 Genetic Engineering
The Southern blotting technique

Some cloning vectors

Cloning vectors	Natural occurrence	Size (Kb)*	Selective marker**
Plasmids			
$_p$ACYC 177	*E.coli*	3.7	Amp^r, Kan^r
$_p$BR 322	*E.coli*	4.0	Amp^r, Tet^r
$_p$BR 324	*E.coli*	8.3	Amp^r,, tet^r. El imm
$_p$MB9	*R.col*	5.8	Tet^r
$_p$RK 646	*E.coli*	3.4	Amp^r

Cloning vectors marker**	Natural occurrence	Size (Kb)*	Selective

Plasmids			
pACYC 177	*E.coli*	3.7	Amp^r, Kan^r
pBR 322	*E.coli*	4.0	Amp^r, Tet^r
pBR 324	*E.coli*	8.3	Amp^r,, tet^r. El imm
pMB9	*R.coli*	5.8	Tet^r
pRK 646	*E.coli*	3.4	Amp^r
pC194	Staphylococcus aureus	3.6	Ery_r
pSA 0501	S.aureus	4.2	Str^r
pBS 161-1	Bacillus subtilis	3.65	Tet^r
pWWO	Pseudomonas putida	117	Kan^r
Cosmids			
pJC 74	Derived plasmid from ColEL	16	Amp^r Ei imm
pJC 720	do	24	Amp^r, Tet^r
pHC 79	Derivative of PBR 322	6	Amp^r, Tet^r
Virusus			
SV40	Mammalian cells	5.2	-
Phage M13$^+$	*E.coli*	6.4	-
Phage *E.coli*		4.9	-

* 1Kb (Kilobase pairs) = 1,000 base pairs = 0.66 mega dalton;

**. Resistance of ampicillin (Ampr), tetracycline (Tetr), erythromycine (Eryr), streptomycine (Sttr), Kanamycin (Kanr), rifampicin (Rifr), and colicin production (EL imm).

The plasmid BR 322 is such an artificially constructed plasmid using two plasmids of *E.coli*-- PBR 318 and PBR 322. This plasmid contains the origin of replication *(ori)* derived from a plasmid related to naturally occurring plasmid Col E1. As a result, the BR 322 plasmid replicates at a rate faster thean bacterial genome. It also possesses genes confering resistance to antibiotics such as ampicillin and tetracyclin. It also has a unique recognition site for 20 restriction enzymes. Certain restriction sites, for example Bam HI in the *tet* gene of the plasmid is present within the gene in such a way that the insertion of foreign DNA will inactivate the *tet* gene. As a result, the recombinant plasmid will allow the cells to grow only in the presence of ampicillin, but will not protect them against tetracyclin. This antibiotic resistance will allow the selective isolation of cells which can be cultured on a medium supplemented with ampicillin.

In addition to the *E.coli* plasmids, a plasmid of gram negative bacteria called RK2 plasmid has been developed which has a very broad host range and has a single restriction site which can be acted upon by several endonucleases.

Isolation of bacterial plasmid

The following steps are ncessary for isolating the bacterial plasmids.

1. Treat the prokaryotic cells with detergents which will release plasmid DNA and chromosome from the bacterial cell along with other molecules. This mixture is known as lysate.
2. Treat the lysate with potassium acetate or acitic acid solution. This precipitates chromosomal DNA and some protein molecule.
3. Remove the precipitate from the lysate by igh speed centrifugation. The supernatant contains plasmid DNA along with RNA, protein and some DNA debris of the chromosome.
4. Treat the lysate with RNAase, this will digest the RNA.
5. The lysate is treated with phenol and as a result two layers are formed – phenol layer and water layer. The water layer contains plasmid DNA, and the other contaminants are found in the phenol layer.

Bacteriophages

A virus which attacks the bacteria is called a bacteriophage. A number of bacteriophages with different genetic materials have been reported so far from a variety of bacteria. Some of these phages are ϕ x 174, phages λ, M13, Fd11, R209 etc. Phages also can be used as cloning vectors like plasmids. Of all the phages, the λ phage perhaps is the one that has been used very frequently as a cloning vector. The λ phage contains a proteinaceous head and a long tail. The head portion has 50 genes in its 49 kb genome of which about half of the genes are essential. On attachment to the host (*E.coli*), the linear DNA molecule is injected into the cell. The linear double stranded molecule of DNA becomes a circle by joining through the single strand of 12 nucleotides commonly known as *cos* site. This *cos* site is a very important feature of λ DNA.

The λ phage has two kinds of life cycles – the lytic phase and the lysogenic phase in the bacterial cell. In the lytic pathway, early in the infection the circular DNA replicates by a rolling circle mechanism and produces long molecules of DNA. At the same time, the phage DNA directs the synthesis of proteins to produce the head. In order to pack the DNA into the head, it undergoes cleavage to yield the fragments of such sizes as to fit in the head. Eventually, a tail is attached to the head. In the lysogenic pathway, the phage genome integrates with the bacterial genome at an attachment site (*att*) where it replicates along with the bacterial genome.

The following are the advantage of using phage as cloning vectors – 1. DNA can be packed *in vitro* into phage particles and transduced into *E.coli* with high efficiency, 2. Foreign DNA upto 25 kb in length can be inserted into phage vector, and 3. Screening and storage of recombinant DNA is easier (Dahl et al., 1981). Before using the λ phage vector, it will be essential to remove the genome from the restriction sites for the enzymes commonly used for cloning. Two types of phage cloning vectors have been constructed – these are the *insertion vector* and the *replacement vector*.

Insertion vector. These have unique cleavage sites into which a relatively small piece of foreign DNA can be inserted. These fragments do not affect the functioning of the phage. The upper and lower limits of DNA that may be packed into the phage particles is between 35–53 kb. The maximum size of a foreign DNA that can be inserted is about 18kb.

Replacement vector. These vectors have cleavage sites present on either side of the length of non-essential DNA. As a result of cleavage, left and right arms are formed and each arm has a terminal *cos* site and a longer stuffer region (the non-essential region), which can be substituted by foreign DNA. The maximum size of foreign DNA that can be inserted into this vector depends on how much the phage DNA is non-essential. It has been estimated that about 25–35kb of genome can be replaced with foreign DNA segments. Example of substituted vectors are–gt, WES and λ 1059.

Cosmids

These may be defined as the hybrid vectors derived from the plasmids which contain *cos* site of λ phage. Reported for the first time by Colins and Hohn (1975), cosmids lack genes for encoding viral proteins. Therefore, neither viral particles are formed, in the host cell nor cell lysis occurs. Special features of cosmids similar to plasmids are origin of replication, a marker gene coding for antibiotic resistance, a special cleavage site for the insertion of foreign DNA, and the small cells. Characters of cosmids dissimilar to plasmids are presence of extraphage DNA, and the *cos* site.

Phasmids

When plasmids are inserted into the genome of λ phage, they are said to be phasmids. The process of insertion of the plasmid DNA into the phage genome occurs in the same way as a phage genome is inserted into the bacterial chromosome. Insertion of plasmid into λ phage is done with a view to have a specific site responsible for recombinational insertion of the phage into bacterial chromosome.

Phasmids may be used in many ways. They can be used as a phage cloning vector from which a recombinant vector may be released.

CLONING ORGANISMS

The vectors that are chosen as vehicles to carry the foreign DNA must be cloned in an organism that multiplies rapidly, thereby producing the gene product in a large quantity. Such organisms into which the vectors are inserted for the purpose of gene expression are known as cloning organisms. A number of microorganisms have been used for the purpose of cloning. The most important among them are the various strains of the colon bacterium, *E.coli.* The strains of *E.coli* used as cloning organisms are HB101, h303 and RR1. The cloning organisms should normally be deficient in DNA recombination as it will reduce undesirable hybridization of DNA. Another essential feature of the cloning organism is it should lack restriction enzymes that might degrade foreign DNA. Strains of bacteria that can bud off mini cells are highly useful if the gene expression is the purpose of experiment (mini cells are those which contain all the elements of gene expression except chromosome).

For a long time, *E.coli* has been the most favourite cloning organism in recombinant DNA studies. Of late, yeasts are also being used as cloning vehicles and they have some advantages over bacteria. They do not live in human beings and are non-pathogenic. Fink et al (1974), for the first time introduced a bacterial gene for the production of leucine into baker's yeast. A snail enzyme was used to induce the yeast to take up bacterial DNA.

Techniques used for identification of clones

Several techniques are employed to identify the isolated clones using a specific molecular probe. Some of these techniques are described below :

1. Polyacrylamide gel electrophoresis

When the genomic DNA extracted from a particular source is digested by a specific restriction enzyme it is broken into several fragments. These fragments of different sizes can be separated through a technique called gel electrophoresis. This separation is necessary to identify a specific gene that is required for use. After these are separated, a specific molecular probe can be used to detect the segments which have sequences similar to those in the probe. For instance, if a gene with a specific nucleotide sequence is required to be separated from a mixture, a probe is prepared with nucleotide sequence complementary to the gene. These (molecular probes) are then allowed to hybridize and the particular gene is identified.

In gel electrophoresis, the DNA molecules migrate from end of the gel to the other. The gel is usually a cylindrical structure, or occasionally even may be a slab. The rate of movement of fragments of DNA on the gel is inversely proportional to the size of the molecules. Heavier fragments will remain at the beginning while lighter fragments will move away. Fragments of different sizes appear as discreet bands on the gel and can be isolated for further study.

Separation of DNA molecules using PFGE (pulsed field gel electrophoresis)
Gel electrophoresis is suitable for only small fragments of DNA which can easily get separated. Large size molecules of DNA, however, cannot be separated using this technique. A new technique called PFGE is used for the separation of large sized DNA fragments, some times representing even whole chromosomes. Using the PFGE technique, separation of DNA molecules belonging to each of the sixteen chromosomes of yeast has been made possible. In this technique, short pulses of electric current are used in two different directions and the DNA is embedded in Agarose plugs to avoid fragmentation. Using this technique, genomes of several fungi have been resolved into chromosomal bands.

Blotting techniques
Gel electrophoresis helps in the separation of fragments of DNA or RNA. The separated bands can be stained and absorbed on the gel. This technique, however, will only help in the separation of fragments of DNA but will not tell as to which fragment contains the desired gene. For this purpose, a molecular probe (which is a DNA fragment) consisting of nucleotides complimentary to the gene are used for hybridization and the required gene is identified. For easy identification and detection, the probe is radioactively labelled. To facilitate hybridization, the bands of the genome are usually transferred to a nitrocellulose membrane in a technique called blotting. Blotting technique was first invented by E.M. Southern, and is known as the southern blotting technique. Two other blotting techniques discovered subsequently are named northern blotting technique and western blotting technique. These are described briefly below.

Southern blotting technique. The DNA sample is first cleaved with a restriction enzyme and the sample is loaded on to gel a electrophoresis for separation. The DNA bands that appear in the gel are denatured into single strands using and placed on a glass plate with its two edges immersed in the buffer. A sheet of nitrocellulose membrane is placed on top of the gel and a stack of many paper towels along with a weight are placed on this membrane (see figure). The buffer solution is drawn up by the filter paper folds and passes through the gel and to the nitrocellulose membrane. Finally, buffer moves up to the paper towels. While passing through the gel the buffer carries with it the fragments of single stranded DNA and binds them on to the nitrocellulose membrane as it passes through it (membrane) to the paper towels above. For the complete migration of the buffer it may take a few hours. Subsequently, the paper towels are discarded, the nitrocellulose membrane with the single stranded fragments of DNA attached to it in different positions is baked at 80ºC to fix the strands permanently on it. This membrane now has the fragments of single stranded DNA fixed on to it in the same position in which they were on the gel. In otherwords, the nitrocellulose membrane is the replica of the gel with its attached DNA fragments. The nitrocellulose membrane is now dipped in a solution containing labelled gene that is present on the membrane. The radio labelled probe hybridizes with the complimentary fragment of DNA on the membrane. The filter subsequently is removed from the solution and washed thoroughly to remove any unbound fragments of DNA. Later, the filter is placed against a photographic film. Wherever the probe has hybridized with the DNA fragments, there will be markings like black dots. This can now be compared with the original gel and the corresponding DNA band can be identified as the required gene.

Northern blotting technique. This is used to identify the required RNA bands. The technique of southern blotting used for DNA transfer from the gel to the nitrocellulose membrane could not be used for transfer of RNA because RNA would not bind with the nitrocellulose membrane. Alwin *et al* (1979), devised a technique in which the mRNA bands from the gel were blot transferred to a chemically reactive paper prepared by diazotization of aminobenzyloxymethyl paper. RNA fragments can bind themselves on to this paper and the paper can be used for hybridization with radio labelled DNA probes. Subsequently, by autoradiography the hybridized bands can be identified. Some of the researches of Thomas (1980), however have shown that mRNA bands can also be blotted directly on to nitrocellulose membrane using special techniques.

Western blotting technique. This technique is used to detect proteins of a particular speciality. Discovered by Towbin *et al* (1979), in this technique newly coded proteins by a transformed cell can be identified. In this method, however, nucleic acid probes are not used. This technique involves the following steps :

(a) Electrophoretic separation of the required protein on a polyacrylamide gel.
(b) Transfering of the separated peptide chains on to a nitrocellulose membrane.
(c) Using radio labelled antibodies as probes the required peptide chain may be hybridized. (The probe consists of antibodies which bind to the antigenic protein).
(d) Autoradiographic identification of the required hybridized protein.

CONSTRUCTION OF CHIMERIC DNA (insertion of DNA into vector)

Chimeric DNA molecules are produced by the insertion of a foreign DNA segment into a DNA molecule of the vector. These chimeric DNA molecules are the means of the cloning genes or other small DNA sequences. Basically, production of chimeric DNA molecules involves breaking open the vector DNA and insertion of foreign DNA followed by joining of the ends. There are three techniques available for the construction of chimeric DNA. These are – pallindromic sequences (cohesive ends) and staggered cleavage, dA-dT tailing and blunt end ligation.

1. Pallindromes and staggered cleavage (cohesive ends). It has already been explained that a pallindromic sequence has complimentary sequences at the two ends of a single strand. Restriction enzymes cutting the pallindromic sequences may have an even cut or a staggered cut. Result of the staggered cut is, DNA fragments with sticky ends are formed. When another DNA molecule (a foreign DNA molecule) is cut similarly, and when the two DNAs are mixed, both DNAs will combine forming a chimeric DNA.

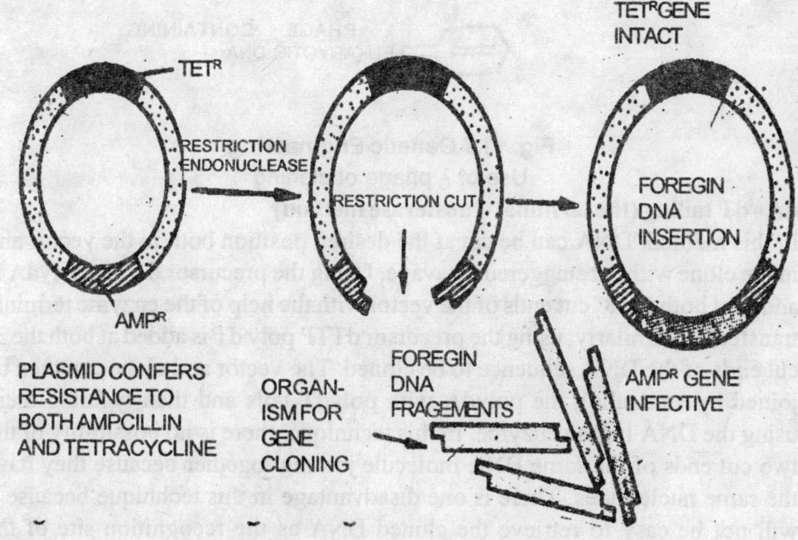

Fig. 11.6 Genetic Engineering

The concept of cloning a foreign DNA in a vector

Discovered for the first time by Boyer and Cohen, this mechanism has an advantage of easily combining two different DNA molecules. Further, it also produces two restriction sites in the chimeric DNA. There are some disadvantages in this method. For instance, the two sticky ends of a vector DNA or of a foreign DNA may themselves join without allowing for the insertion of one into the other.

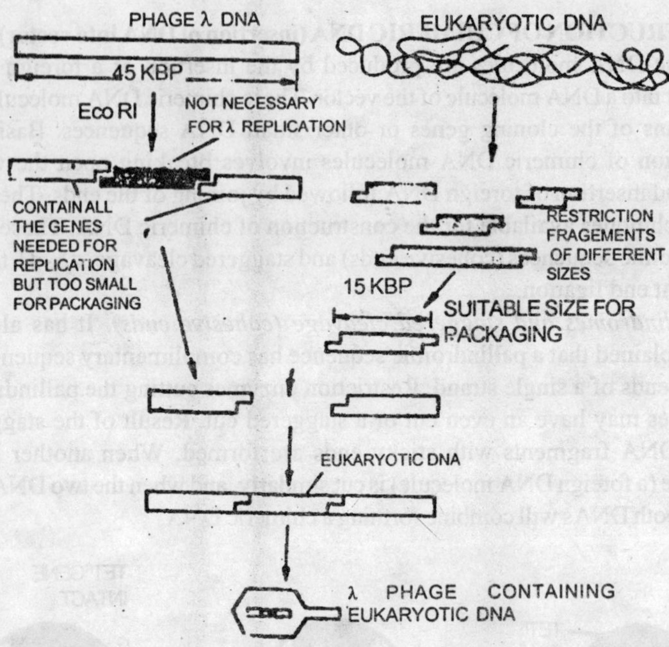

Fig. 11.7 Genetic Engineering
Use of λ phage of cloning

dA - dT tailing (the terminal transferase method)

In this method, DNA can be cut at the desired position both in the vector and in the clone without staggered cleavage. Using the precursor dATP, polydA is added at both the 3' cut ends of the vector with the help of the enzyme terminal transferase. Similarly, using the precursor dTTP poly dT is added at both the 3' cut ends of the DNA sequence to be cloned. The vector and clone can then be joined by anniealing the polydA with polydT tails and then lygating them using the DNA lygase enzyme. In this technique, there is no possibility of the two cut ends of the same DNA molecule joining together because they have the same nucleotides. There is one disadvantage in this technique because it will not be easy to retrieve the cloned DNA as the recognition site of the enzyme is lost due to the addition of poly dA and poly dT. It has been found out that the use of poly dG and poly dC will help in the restoration of the recognition site.

Blunt end ligation

In this technique, a restriction enzyme is used to cut a duplex DNA at the same site in both the strands. The broken ends are then used for joining with the two

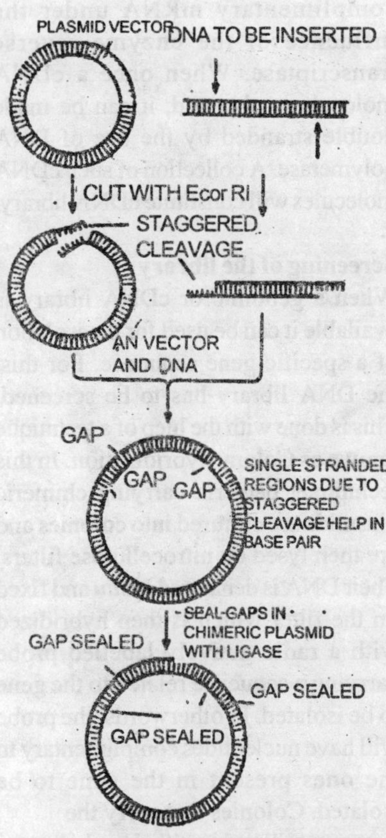

Fig. 11.8 Genetic Engineering Cloning of a segment of DNA staggered cleavage

ends of foreign DNA molecule irrespective of the sequences present at the broken ends of the two molecules. The blunt end ligaton is made possible by the use of the T4 DNA lygase. This technique also has a disadvantage in that the broken ends of the same DNA molecule may join together.

Construction of genomic and cDNA libraries

In order to isolate required gene or genes from a genom a mixture of clones each carrying DNA fragments derived either from the natural genome or from cDNA is prepared. This mixture contains thousands of clones and is collectively called genomic library. Similarly, when cDNA clones are collected it is known as cDNA library.

Construction of genomic library

Cloning an entire genome in the form of fragments mixed randonly is known as genomic library. This may be done by a method known as shotgun method (see earlier in the same chapter for details). With a genome library, clones can be perpetuated indefinitely in a plasmid vector and retrieved when ever necessary.

Genomic libraries can also be prepared by using a number of restriction endonucleases one at a time so that one will have fragments of different sizes with cuts at different places. This has a disadvantage in that many a time the incision site many be within a gene so that complete genes may be difficult to obtain. This may be overcome by using restriction enzymes which have very short recognition sequences so that the fragments that are obtained will be long and the chances of slicing within a gene are avoided.

cDNA library

It has already been explained above as to how to obtain a cDNA using its

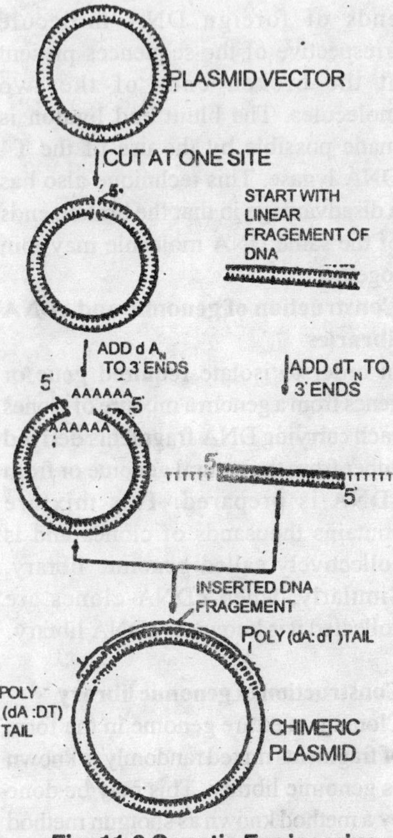

Fig. 11.9 Genetic Engineering Cloning of DNA segment by dA-dT tailing

complimentary mRNA under the influence of the enzyme reverse transcriptase. When once a cDNA molecule is obtained, it can be made double stranded by the use of DNA polymerase. A collection of such cDNA molecules will constitute cDNA library.

Screening of the library

When a genomic or cDNA library is available it can be used for the isolation of a specific gene sequence. For this, the DNA library has to be screened. This is done with the hlep of a technique known as Colony hybridization. In this technique, bacteria carrying chimeric plasmids are cultured into colonies and are then lysed on nitrocellulose filters. Their DNA is denatured *in situ* and fixed on the filter which is then hybridized with a radio actively labelled probe carrying a sequence related to the gene to be isolated. In otherwords, the probe will have nucleotides complimentary to the ones present in the gene to be isolated. Colonies that carry the sequence will be identified by dark spots after radioautography so that the colonies carrying the vector with the desired gene sequence can be recovered from the master plage and used for further experiment.

Transfer of the recombinant DNA into a cloning organism

When once a mixture of recombinant DNA is obtained it has to be transferred to a suitable cloning organism for the purpose of expression. Several techniques are used to transfer the recombinant DNA into suitable cloning cells. Transforming the cloning cells with recombinant DNA is a difficult process, as the cells of bacteria *(E.coli)* have restriction enzymes which will immediately degrade the foreign DNA. To escape from degradation, exponentially growing culture of *E.coli* is pretreated with $CaCl_2$ at low temperature and thereafter the DNA is mixed up. This treatment is known to sufficiently prevent the action of restriction enzymes. When the recombinant plasmids get established in the cells of *E.coli*, the cells are said to be transformed.

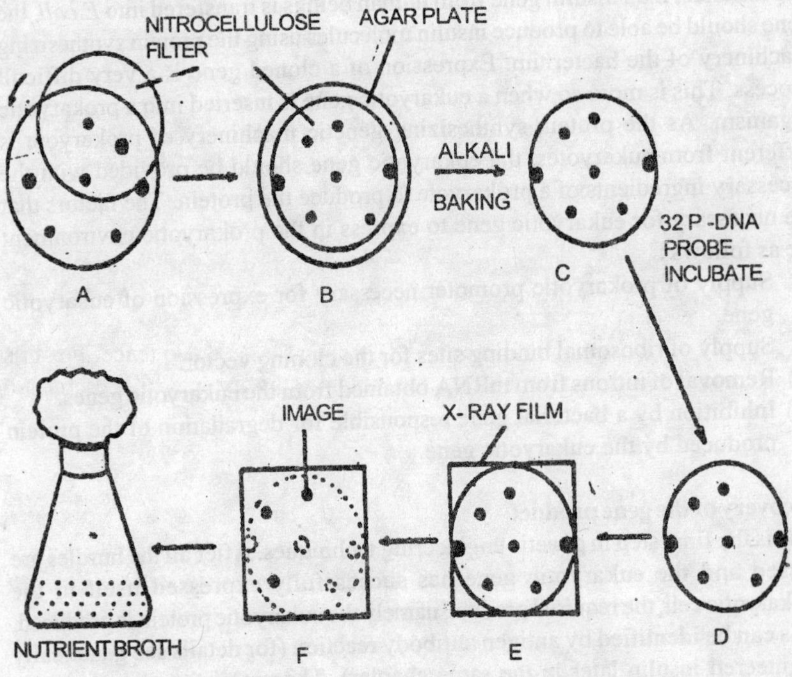

Fig. 11.10 Genetic Engineering
Colony hybridization technique

Instead of plasmids even, λ phages could be used as transforming elements. The entry of the phage into *E.coli* is usually termed transfection (a combination of transformation and information).

Not all the cells will accept recombinant DNA and get transformed. It has been estimated that about 10^5 transformats per microgram of cloned circular plasmid can be generated. The cells which have been transformed will grow into colonies along with other non-transformed ones. Colonies with transformed cells can be identified with suitable techniques.

Expression of cloned DNA

The purpose of genetic engineering is not just to separate and clone the genes in a suitable organism. Isolation and multiplication of genes is only a means ultimately to obtain the gene product namely the required protein. Hence, the recombinant DNA molecules should be allowed to express in the new environment (*E.coli* cells), and produce what ever protein they were producing. For instance, if an insulin gene from human beings is transfered into *E.coli,* the gene should be able to produce insulin molecules using the protein synthesizing machinery of the bacterium. Expression of a cloned gene is a very difficult process. This is more so when a eukaryotic gene is inserted into a prokaryotic organism. As the protein synthesizing genetic machinery of prokaryote is different from eukaryotes, the eukaryotic gene should be provided with the necessary ingredients of a prokaryote to produce the protein. The factors that are necessary for eukaryotic gene to express in the prokaryotic environment are as follows :

(i) Supply of prokaryotic promoter necessary for expression of eukaryotic gene.
(ii) Supply of ribosomal binding sites for the cloning vector.
(iii) Removal of nitrons from mRNA obtained from the eukaryotic genes.
(iv) Inhibition by a bacterial gene responsible for degradation of the protein produced by the eukaryotic gene.

Recovery of the gene product

This is the final step in genetic engineering techniques. After all the hurdles are passed and the eukaryotic gene has successfully expressed itself in the prokaryotic cell, the required product, namely the eukaryotic protein is produced. This can be identified by antigen antibody reaction (for details see genetically engineered insulin later in the same chapter). The protein produced can be purified by standard chemical processes.

APPLICATIONS OF GENETIC ENGINEERING

Scope of genetic engineering (Biotechnology)

The possibilities are immense and the sky seems to be the limit for future biotechnologists. The following are some of the applications of genetic engineering.

I. Gene therapy

A new branch of curative medicine is opened up due to genetic engineering. Many genetic disorders like haemophilia, alkaptoneuria, galactosemia etc., which are due to defective genes can be cured by replacing them (defective genes) with correct genes. Plasmid or virus mediated genes may be incorporated into human systems.

II. Cheap production of drugss

Many of the drugs like vitamins, antibiotics, harmones are now being produced chemically or produced from higher organisms which is a slow and costly process. If the genes for the production of these are introduced into rapidly multiplying bacterial cells, we have virtually myriads of tiny living factories turning out the medicines at a fraction of their original production cost.

III. Gene for new diet

Some of the plants or plant products are presently non-edible for want of enzymes in the human system to digest them. For ex, wood, grass, etc. If the genes that code for the digestive enzymes (for materials like wood etc) could be introduced into the human system, wherein there would virtually be no famine on earth. Another possibility, (which sounds unbelivable at present) could be to introduce genes that can produce the chlorophyll pigment in the animal or human system, wherein we would be solving the food problem once for all !.

IV. Role of genetic engineering in Agriculture

The following are some of the advantages of gene recombination in crop improvement.

(a) Production of new varieties

New varieties of plants that are not present in nature can be produced by the introduction of new genes.

(b) Somatic hybridization

Two crop plants with important characters, which could not be hybridized sexually due to genome incompatibility can be hybridized through protoplast fusion. Production of *Pomato* (Tomato x Potato) is one such example.

(c) Increasing photosynthetic efficiency

Ribulose diphosphate carboxylase (Ribisco) is a key enzyme in the dark fixation of CO_2. Modification of *Rubisco* to reduce photorespiration (Osmond 1981) by elimination of oxygenase activity would increase-the CO_2 fixation rate. The enzyme has two sub units and is known to be coded by seven genes. If these genes could be altered, resulting in the changed *Rubisco,* it will increase the photosynthetic efficiency.

(d) Nitrogen fixation

The roots of leguminous plants normally possess nitrogen fixing nodules, induced by various strains of *Rhizobium.*m As a result, leguminous plants require little or no nitrogen fertilizer. Aerable crops require about 100 kg of nitrogen per hectare. If this can be replaced by N_2 fixing genes, it would lead to a lot of saving in terms of fertilizer cost. Nitrogen fixation is catalyzed by the ATP dependent electron reduction of dinitrogen to ammonia by nitrogenase. If it is possible to introduce the gene to higher plants, it would bring about a revolution in agriculture. This may be achieved by one of the following ways.

1. Transfer of nitrogen fixing gene (*nif*) from the living bacteria like *Klebsiella, Azatabacter* etc., to higher plants. This would enable the plants to fix the N_2 directly. This is still in the experimental stage, as the *nitrogenase,* produced by the *nif* gene is highly oxygen liable and gets degraded in higher organisms.

2. Expansion of host range of symbiosis to crops other than legumes. At present non leguminous plants do not have symbiosis with *Rhizobium.* If it is possible to expand the host range for the bacteria, a lot more plants will have the capacity of N_2 fixation.

3. Increasing the efficiency of symbiotic bacteria to N_2 fixation. By gene manipulation, it may be possible to increase the N_2 fixing capacity of the symbiotic bacteria, which will benefit the plants.

(e) Herbicide resistance, disease resistance etc.

By gene manipulation, it will be possible to introduce this character into plants.

In conclusion, it may be said, that genetic engineering is a highly potential tool by which man can alter the macromolecule of inheritance, and thereby change the biological environment to his choice. But the new technique has to be used judiciously, otherwise it can also create untold harm to human society, if unscrupulous scientists use it to create destructive genes. A few of the specific applications of genetic engineering are discussed below in some detail.

Nitrogen fixing genes (*nif* genes)

Nodule formation and nitrogen fixation are two of the important biological processes which are genetically controlled in diazotrophic bacteria. From the genome of *Rhizobium,* several genes coding for nodulation (*nod* genes), and nitrogen fixation (*nif* genes) have been identified and isolated. In free living and symbiotic nitrogen fixers, nodule forming and non-nodule forming *nif* genes are present either on the genome or on the plasmids in their cells.

Nod genes

Nodule formign species of *Rhizobium* contain in their cells an extremely large plasmid called *mega plasmid,* which consists of a number of *nod* genes and *nif* genes (Rosenberg et al, 1981). Both the genes are closely located.

Kondorosi *et al* (1982), were able to transfer the megaplasmids from one species of *Rhizobium* to the other simultaneously transfering the ability of nodule formation.

The nod gene region is of 8.5 kb in nucleotide sequence length, and proteins encoded by the 8.5 kb fragment when transferred by *E.coli* has been determined. The *nod* gene is a cluster made up of 4 genes designated as *nod a, b, c* and *d.* A comparison of *nod* genes between different species of *Rhibozium* shows that about 70% of the nucleotides are homologous.

Nif genes

The *nif* genes are present in several species of Rhibozium and are located on a megaplasmid adjacent to the *nod* genes. *Nif* genes have also been identified in many cyanobacteria such as *Anabena,* where the *nif* genes are located on the main genome.

Early studies on *nif* genes were carried out in *Klebsiella pneumoniae,* and the function of these genes was confirmed by transfering into *E.coli* cells. Researches have also been conducted to identify the product of *nif* gene cluster. The *nif* gene cluster consists of 7 operons.

In cyanobacteria, *nif* genes have been located both in vegetative cells as well as in heterocysts. But there is some difference in the nature of the *nif* gene between the vegetative cells and the heterocysts.

Cloning of nif genes

Researches are being conducted in many countries to transfer the *nif* genes to higher plants especially to monocots. If this is successful it will be indeed a great boon to agriculture as there will be plenty of saving in nitrogen fertilizers, for the plants themselves will be able to fix atmospheric nitrogen. Gene expression of *nif* genes in eukaryotic cells is very difficult because the nitrogenous enzyme required for fixing the nitrogen and controlled by the *nif* gene would get denatured in the presence of oxygen. If this can be overcome, *nif* genes can be easily transferred to eukaryotes. *Nif* genes, however, have been cloned in *E.coli.* One of the methods of transfering *nif* genes is to transfer them into the chloroplast of eukaryotic cells as the translational and transcriptional processes of chloroplasts resemble prokaryotic features. These attempts might succeed because the chloroplasts are able to produce the ATP and reducing power both of which are required for nitrogen fixation (Merrick and Dixon, 1984). But there are major problems for such a transfer of *nif* gene to chloroplasts. These are:

1. Lack of chloroplast transfering technique.
2. Protection of nitrogenase from oxygen evolved during photosynthesis.

Genetically engineered Insulin

Insulin is an important hormone that plays a vital role in the metabolism of carbohydrates. In patients suffering from diabetes, insulin has to be injected subcutaneously to regulate the blood sugar level. Insulin producing genes can be transferred to bacteria by techniques of genetic engineering and large quantities of insulin may be extracted. This process is cheaper and convenient than the conventional one. We will briefly discuss below the method of procuring insulin by genetic engineering techniques.

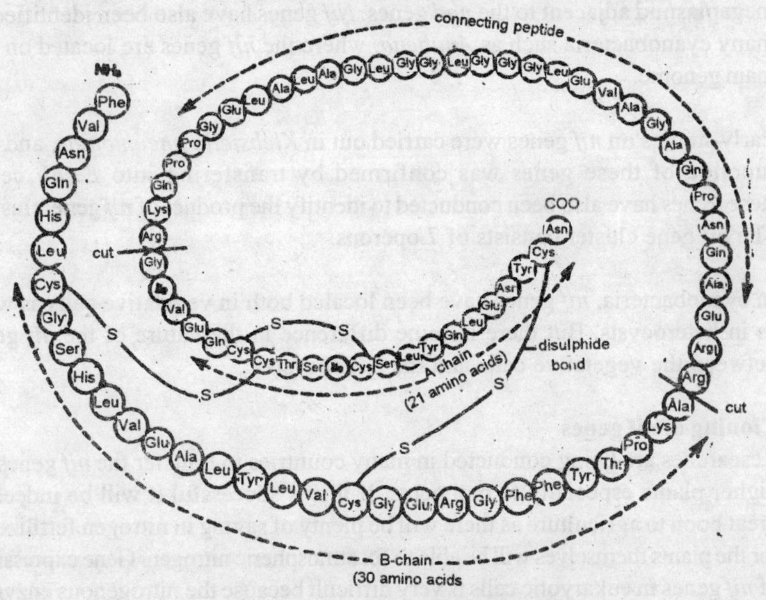

Fig. 11.11 Genetic Engineering
Structure of Insulin molecule

Structure of human insulin

Insulin molecule is composed of two chains – α and β chains of aminoacids. Initially, it is produced as a single long chain called *Pre-proinsulin* which is later cut into a shorter chain called *proinsulin* (composed of 84 aminoacid units). Proinsulin then breaks into α and β chains when a piece from the middle is cut off. Chain α is composed of 21 aminoacids and chain β has 30 amino acids. The remaining 33 aminoacids act as a connecting peptide chain.

Production of Insulin in E.coli The genes for insulin chain α and β are identified and located. This may be done first by identifying the codons and then the corresponding mRNAs can be reconstructed. Using these mRNA molecules, and with the help of the enzyme *reverse transcriptase,* the corresponding segments of double stranded DNA molecules are produced. Then with the help of the enzyme *terminal transferase,* the ends of the insulin genes (DNA strands) are extended with short sequence of identical bases.

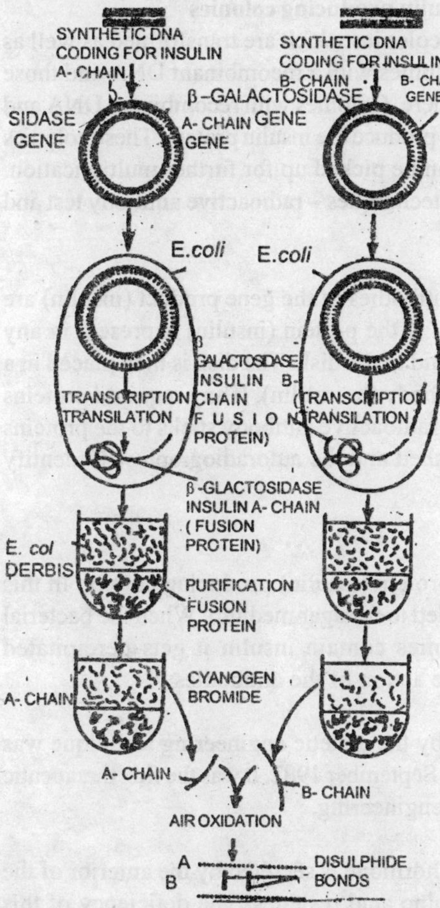

Fig. 11.12 Genetic Engineering
Technique of producing genetically engineered insulin in E.coli

Plasmids obtained from *E.coli* are opened up at specific portions by the use of restriction enzyme. The insulin genes (DNA strands) of α and β chains are added to the plasmids separately forming recombinant DNA plasmids. Gaps if any in the circular plasmids of α and β chains are filled by lac operon gene for switching on the synthetic process. The plasmids with recombinant DNA for α and β chains are inserted into *E.coli* finally.

The recombinant DNA plasmids multiply in *E.coli* and trigger the synthesis of α and β chains of insulin as a genetic expression. Following large scale production cultures, the insulin chains are obtained following specific purification processes. Finally the α and β chains are chemically bonded by disulphide bonds to produce the complete molecule of insulin.

Gilbert and Villakomaroff (1980), isolated mRNA for insulin from β cells of the pancrease of rat and inserted it into PBR 322 plasmid in the middle of a gene normally coding for the enzyme penicillinase and incorporated it into the cloning organism (*E.coli*), *E.coli* cells produced the hybrid protein consisting of penicillinase + proinsulin. Proinsulin was then separated from penicillinase.

Immunological tests to identify insulin producing colonies

On a culture medium, there will be colonies which are transformed as well as non transformed. In otherwords, colonies with a recombinant DNA and those without recombinant DNA will be there. Colonies with recombinant DNA and successfully expressing the gene will produce the insulin protein. These colonies have to be identified so that they can be picked up for further multiplication. This is done by two immunological techniques – radioactive antibody test and immunoprecipitation test.

Radio active antibody test

in this test a plastic disk to which antibodies of the gene product (insulin) are bound is pressed on to the colonies. If the protein (insulin) is present in any colony, it will bind to the antibody and to the disk. The disk is then placed in a solution containing radio active antibody (to insulin). If any required proteins (insulin) are present on the disk, the radioactive antibody sticks to the proteins on to the disk. The disk is then washed and the autoradiograph will identify colonies if any producing insulin.

Immunoprecipitation test

This test is also used to identify a protein (insulin) producing colony. In this technique, insulin antibodies are added to the agar medium. When the bacterial colonies grow, if any of the colonies contain insulin it gets percipitated surrounding the colonies, due to the action of the antibodies.

Human insulin Humulin produced by the genetic engineering technique was made available to Diabetic patients in September 1982. It was the first therapeutic product to be produced by genetic engineering.

Human growth hormone (HGH)

Also known as Somatotrophin, this hormone is secreted by the anterior of the pituitary and is made up of 190 amino acid residues. The deficiency of this hormone results in dwarfism. For the conventional treatment of dwarfism, HGH is injected into the patients. But the extraction of hormone from the pituitary glands cannot meet the ever increasing demand for this hormone. Genetic engineering procedures have successfully produced somatotrophin.

Double stranded cDNAs were obtained from the mRNA precursors of HGH and then incorporated into bacterial cells where they expressed in non-precursor form. The bacteria were unable to convert the peptides of HGH into biological active forms. A recombinant plasmid containing a full length of HGH cDNA is cleaved with restriction enzymes that release fragments containing the complete

HGH coding sequence after codon 24. Several overlapping complimentary oligo nucleotides were joined to form a synthetic strand of a small DNA fragment which contains the coding sequence for the first 24 aminoacids of mature HGH. The synthetic DNA and cDNA molecules are joined together to produce a new fragment which has the complete coding sequence of HGH. This is then ligated to a restriction site a little below the lack promoter operator gene cloned on a plasmid. These plasmids are cloned in *E.coli*. Synthesis of HGH is induced by an inducer (IPTG) of lack operon. The HGH thus obtained is later purified by chemical methods.

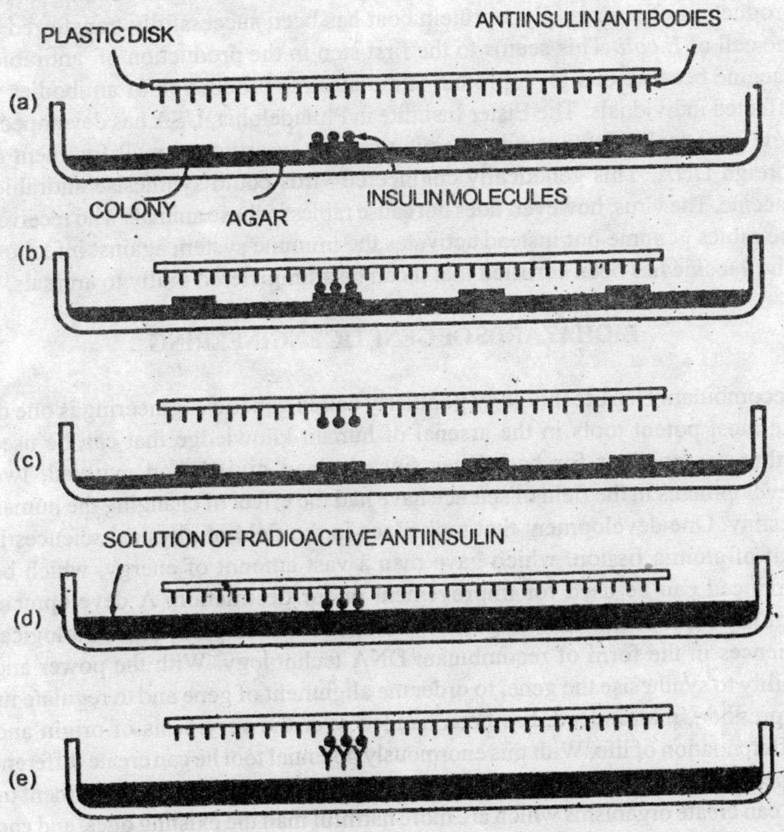

Fig. 11.13 Genetic Engineering
Radioactive antibody test

About 1 lakh molecules of the hormone per cell of *E.coli* have been produced by genetic engineering techniques. One of the problems in this is the attachment of an extra methionine to the molecule. This has to be removed to make the molecule biologicaly active.

Vaccine for Rabies virus (RV)

Hydrophobia caused by mad dog bite used to take a heavy toll of the human life. The credit of discovering a method to cure hydrophobia or rabies goes to Louis Pasteur. Even now however, due to ignorance and not taking the vaccine in time, is causing a number of death of humans and animals in many Asian and African countries. The conventional antivirus vaccine is difficult to obtain and also a tedious process. Researches are being conducted to produce genetically engineered rabies vaccine using *E.coli.* Attempts are made to isolate the mRNA coding for the viral protein from the RV infected cells. The genes coding for the production of a virus, glycoprotein coat has been successfully transferred to the cell of *E.coli.* This seems to the first step in the production of antirabies vaccine because this glycoprotein stimulates the production of antibodies in affected individuals. The Eister Institute in Philadelphia, USA has developed a new genetically engineered vaccinia virus by inserting a small fragment of foreign DNA. This genetically engineered virus could synthesise antirabies vaccine. The virus, however, does not cause rabies in those animals who received the rabies genome but instead activates the immune system against infection. The vaccine has been certified that it can be administered orally to animals.

BIOHAZARDS OF GENETIC ENGINEERING

Recombinant DNA technology popularly called genetic engineering is one of the most potent tools in the arsenal of human knowledge that can be used either for good or for bad. Ever since human civilisation evolved, two developments in the field of science have had the effect of changing the human destiny. One development that took place in the field of physical sciences is that of atomic fission, which gave man a vast amount of energy, which be produced can be used for improvement or for destruction. A development similar to this or much more potent than this has taken place in the biological sciences in the form of recombinant DNA technology. With the power and ability to synthesise the gene, to order the alignment of gene and to regulate its expression, man indeed, has been handed over the very basis of origin and differentiation of life. With this enormously potential tool he can create different types of organisms, vastly modify the existing organisms for his betterment or he can create organisms which are more harmful than the existing ones, and can even spell the deathknell of the very nature which has created him.

The brief discussion above will clearly point out to the fact that research in genetic engineering has to be carefully controlled, manipulated, and should be encouraged only when it is absolutely necessary, and that too with plenty of safeguards. In the paragraphs below, we will briefly outline the biohazards of uncontrolled genetic engineering research and what safeguards have to be followed while following the techniques of recombinant DNA.

Harmful aspects

Genetic engineering involves introducing a foreign DNA into another organism so that it gets transformed. Many a time, when a gene segment is transferred one may be aware of its expression in the original organism. But one is not too sure as to what kind of expression it may have in the new environment i.e., in the new organism. This kind of recombination many a time creates a dangerous new form either deliberately or accidentally. A harmless microorganism might turn out to be a very harmful one after it acquires new genes from another organism. Eventhough it is true that the genes introduced by themselves may not be harmful in their original combination, but when they are introduced into a new set up they might bring about the potentially dangerous expression. If, by chance, a transformed organism comes to carry a deadly disease due to genetic manipulation and if that organism escapes out of laboratory, one can imagine the uncontrolled horror that the organism might bring about. The genetically manipulated organism might be so potent in its pathogenicity that any of the existing drugs may not be able to combat it. For instance, *Streptococcus,* a pathogenic organism was never resistant to penicilein originally in nature. For the purpose of genetic engineering research, a plasmid with a desired gene is introduced into *Streptococcus* and at the same time to identify the transformed organisms, a marker gene such as penicillin resistant gene is itnroduced, when the organism becomes penicillin resistant the antibiotic now would be useless. If such an organism even though created in the laboratory were to escape one can imagine the dangers caused by it.

Escape of transformed microbes into environment

Inspite of precautions being taken, their always exists a possibility that a recombnant orgnanism might escape from laboratory and get permanently entrenched in the ecological system. If the organism in question were to be a parasite, it would cause immense havoc to all those organisms it parasitizes. Microorganisms that live in special ecological niche like hot spring or salt water might not pose such a great problem as they cannot spread widely.

The escape of microorganisms from the laboratory may be by design or by accident. If it is by design, then one has to understand that microbes are allowed to escape deliberately to cause harm to the environment outside. Accidental escape of chimeric microbes from the experimental laboratories may take any one of the following routes.

(1) Laboratory personnel such as scientists, attendants, visitors, sweepers etc. might inadvertently carry the microbes and spread them outside.

(2) Microbes might also find their way out through laboratory facilities such as glassware, drainage of chemicals, spillage of media, and such other inanimate objects.

In addition to the above, non-scientific personnel in the laboratory many a time may not be aware of the stringent precautions that have to be mandatorily taken while dealing with pathogenic microbes. Many a time, even if they are aware, they might just not follow the precautions as they are quite tedious and and cumbersome. Laboratory personnel while constantly dealing with pathogenic microbes many a time might acquire immunity and may not suffer from the consequences of contact with pathogens, but they act as carriers and others who come in contact with them will have to suffer the consequences.

Unintended escape of microbes and possible creation of new microbes possesses perhaps the greatest danger to the environment. In this context, it is worthwhile mentioning the opinion of three international experts on AIDS – Dr.John Seale from London, Dr.Robert Strecker from California, and Dr.Jacob Segal from Berlin. These scientists claimed in 1987 that the AIDS virus was artificially created by American scientists during laboratory experiments when they were trying to discover a cure for cancer by combining parts of Maedy - visna virus found in the sheep and human T-cell Leukaemia virus type 1. They argued that the scientists or other personnel in the laboratory could have been infected with the new virus through a minute cut or even by inhaling (Ignacimuthu, 1995).

Plus and Minus points of genetic engineering research

A great controversy is going on currently in the International Community involving biologists, social scientists, politicians and religious people regarding the desirability or otherwise of conducting genetic engineering research. While many people argue that genetic engineering research would be a harbinger of prosperity, others are equally vehement in opposing that research in genetic engineering would cause untold harm to human life as it is trying to tinker with the very basis of life and evolutionary process. The arguments for and against recombinant DNA technology are summarised below.

Arguments favouring genetic engineering research

1. The transfer of genetic material from eukaryotes to prokaryotes, and *vice versa* have been going on even in nature perhaps on a very small scale. If

nature were to intend to prevent the transfer of a genetic material, evolution would have surely brought in a perfect genetic barrier preventing such a transfer. In fact, ever since the origin of life, evolution could proceed only because nature took to genetic engineering.

2. The dangers of genetic engineering research perhaps will recede with time as people and scientists become experienced and accumulate knowledge.

3. Enough safe guards are already available in research laboratories dealing with pathogenic organisms. They can be made much more stringent. In any case, the beneficial aspects of combating the disease perhaps far more outweigh the possible dangers.

4. Genetically transformed microbes have been designed with a special purpose. Hence, the argument that they would compete with the other normal microbes and perhaps might be picked by natural selection appears to be imaginary. Further, with attenuated characters, the transformed organisms will not be able to survive unless special conditions such as the laboratory environment are provided with.

5. The idea that by acquicuring a new gene the microorganisms become pathogenic is not true, because in any genetic transformation the host organism will always reject a virulent gene combination and the unwanted genes are generally lysed by restriction enzymes and even if the genes express themselves, the gene product, namely the foreign protein gets degraded by proteolytic enzymes.
6. All types of genetic engineering research need not be dubbed as harmful and dangerous. For instance, research in the field of agriculture, dairy and others will not pose any problem. Researches in the medical field, however, can be continued with stringent regulations and control. At the same time however, one cannot simply push aside the vast benefits that have accrued to mankind because of improved health care products from genetic engineering. Hence, total stoppage of genetic engineering research would weaken the ability of man to understand and control nature and ultimately it is not in the interest of man himself.

Arguments against genetic engineering research

1. Unintended or accidental escape of microbes from the laboratory is possible through many routes. This escape of the transformed organism into the environment will pose great danger.

2. Genetic engineering research actually tinkers with the very basis of the

evolutionary process. Nature indeed has been a very efficient genetic engineer and it has been doing this work very slowly and steadily over a period, of millions of years. Transfer of genetic material and production of new traits in nature takes millions of years to be achieved. If this is sought to be achieved in a very short period it amounts to short circuiting the normal process and the consequences could be really hazardous. When an individual is fashioned out of natural process, it will be a package of both weakness and strength. If we were to tinker with this process it would naturally disturb the fine tuned balance and relationship that is existing between the organisms since millions of years.

3. The intended benefits of genetic engineering research can be actually achieved through other research methodologists. To achieve these goals one need not resort to such dangerous techniques like genetic engineering. The benefit of genetic engineering research does not justify the cost (not just in terms of money) involved.

4. With the kind of diverse political set up prevailing in different countries of the world where dictators, fundamentalists still have a very powerful say, this technique can definitely be misused and there is every possibility that these people might unleash a biological warfare on countries whom they consider inimical to them. For instance, water bodies of a country could be infected with new versions of pathogens or their environment might be sprayed with pathogenic microbes. All these possibilities should make one think whether it is worthwhile indulging in this kind of research.

Safe guards in genetic engineering research

The scientific committee at large and biologists in general, have been thinking for a very long time as to how to introduce some sense into genetic engineering research. While it may not be prudent to indulge in genetic engineering for problems which can be solved otherwise, it is equally true that genetic engineering cannot be totally given up as it indeed has the potential of solving some of the serious problems confronting the human race, which other research methodologies cannot hope to solve. The only answer to this being, research in genetic engineering should be properly regulated, monitored and if necessary with stringent international control.

A committee headed by Paulberg (1973-74), an eminent scientist, held discussions with other interested persons and published a small bulletin outlining the risks involved in genetic engineering research. In 1975, an International group of scientists who were very much concerned with genetic engineering research convened a meeting in California, USA and considered the ethical and safety implications of genetic engineering research. An outcome of this meeting was the formulation of a series of guidelines that sought to regulate and in some cases even prohibit research in recombinant DNA. In the USA, the National Institute of Health (NIH) issued guidelines in the year 1976 for US scientists involved in genetic engineering research. The guidelines required among other things that any research proposal involving recombinant DNA should be submitted to a special committee of NIH for approval. NIH also graded the laboratories into four levels (P1 to P4) where genetic engineering research of different levels were recommended. For instance, experiments involving DNA from primates and animal viruses involving harmful genes could be carried out only in extra secure P4 laboratories. In 1982, an advisory panel of NIH - the recombinant DNA advisory committee (RAC) published revised guidelines. According to these new regulations, recombinant DNA experiments were graded into 5 classes, and of these it is essential that three levels of research must obtain NIH approval even before initiation. These include researches involving drug resistance of microorganisms, toxin synthesizing genes and operations involving release of genetically engineered organisms into environment (as those used for pest control).

On the 26th January, 1982, a meeting was held of the parlimentary assembly of the council of Europe in Strasbury and recommended certain guidelines aimed at protecting the individual from the inheritant dangers of genetic manipulation. Three areas were mentioned in particular – genome map of human cells, gene therapy of somatic cells and gene therapy of germ cells. The report of the council concluded that inspite of the possible advantages of genetic engineering research, man made decisions should not substitute the free play of nature and everyone's right not to be genetically manipulated should be affirmed (Ignacimuthu, 1995).

Outside USA, the regulations on genetic engineering research are actually as stringent as they ought to be. France and Germany are considering whether to restrict research in genetic engineering because of the ethical questions involved. Recombinant DNA guidelines of France are ambiguous about the deliberate release of genetically altered organisms. In Germany, however, release into environment of genetically altered organisms are strictly prohibited. Many social activist groups are opposing gene therapy and such other genetic engineering parctices in terms of ethical and moral parameters.

In India, genetic engineering research is still in infacy, yet considering the importance of it, Government of India has constituted four committees for this purpose.

1. Recombinant DNA Advisory Committee (RDAC). To evolve guidelines.

2. Institutional Biosafety Committee (IBSC). For implementation of guidelines.
3. Review Committee of Genetic manipulation (RCGM). To recommended special suitable conditions for experiments and trial.
4. Genetic Engineering approval committee (GEAC). For large scale manufacturing, environmental release of genetically altered organisms/ products and import/export of genetically altered organisms.

The followng may be considered as some of the safeguards that should involve any research in genetic engineering.

1. Harmful genes, such as those coding for toxins and antibiotics should not be cloned without extreme precautions.

2. Animal genes, genes of tumour viruses should not ordinarily be cloned in bacteria or others microbes.

3. Laboratory facilities should be such that there should be virtually no possibility of the escape of genetically altered microbe. In addition to scientists, even the other ordinarily personnel of the laboratory should be thoroughly briefed about the dangers of carelessness before they are employed. Workshops on precautionary measures should be conducted periodically so that no one violates them. There should be mandatory check of laboratory equipments of cabinets, hoods and other glassware to see that they are microfree when they are taken out of the laboratory. Laboratory should also possess special equipments like negative pressure rooms, safety cabinets, special traps and drainage etc., to prevent the possible escape of the microbe.

4. Use of microbes which can live only in special environmental conditions should be picked up for genetic engineering research, because even if these organisms escape into the environment after genetic alteration, they will not survive in the normal environmental conditions so that the chances of their rapid spread are very less.

5. As far as possible, cloning vectors should be non-conjugative plasmids because even if they are uncontrolled they will not be able to promote their own transfer by conjugation.

6. The cloned vector should be sufficiently modified are altered in such a way that it will not be able to express itself in any environment, but can do so only under special circumstances like a particular host strain that are normally not available The genetic alteration must be such that in case it is transferred to another microbe, it should prove lethal to it.

7. The cloning organisms should be modified in such a way that after cloning, they depend upon very special conditions of existence which can be provided only in a laboratory and are not available in the environment.

8. Attempts are being made to produce "crippled microbes" for use in genetic engineering research. The genome of these microbes is altered and disabled in such a way that if they escape from the special laboratory conditions they will not survive. The following are some of the instances of crippled microbes for use in genetic engineering research.

Diaminopimelic acid (DAP), is an important component of the cell membrane in *E.coli*. Genetically altered *E.coli* were produced which were DAP deficient. These cannot survive unless supplemented by DAP. Hence, these organisms even if they escape outside cannot survive. Such organisms are ideal tools for genetic manipulation. Precautions are, however, necessary because many a time microbes are known to undergo mutation very fast. In fact in the above example, the DAP deficient *E.coli* was very soon mutated back and was able to produce DAP. Scientists however, discovered that there was more then one gene involved in the production of DAP and deleted that also. But the bacteria would not yield so easily to the manipulation of the microbiologists. The DAP deficient bacteria again survived by producing a sticky substance called Calanic acid around their protoplast. But then, the calanic producing gene was again removed. This will clearly reveal that our knowledge on the genome of microbes would always remain incomplete because we will be able to identify a gene only when it expresses itself after natural selection, for instance calanic acid producing gene was not selected to express when DAP producing gene is there. In the absence of the latter, the former was selected. Who knows in the absence of calanic acid producing gene which other bacterial gene which is hybernating may get selected and express itself. After all life is a struggle and no individual will give up its existence so easily.

Genetic engineering research and ethics

Man is a product of evolution and, however much he may think that he is different from other organisms because of his intelligence, the stark truth is that he is definitely a part of biological spectrum whether he likes it or not. Man is neither the first step nor the last step in the evolution of life. The process of evolution which has produced so many diverse types of organisms has also produced man and all these organisms that form the biological spectrum were not produced within a day. Even the evolution of human beings took several millions of years. It is quite true that the millions of organisms in the present day biological spectrum are produced out of nature genetic engineering. But one must realise that these have been crafted painstakingly and patiently over the course of millions of years. Man's attempt of fast genetic engineering seeks to destabilise the delicate balance that has been inbuilt into the relationship of all organisms. Further, the diversity of life and its glorious uncertanity is the spice that makes life meaningful. Present day world consists of beneficial organisms, harmful organisms, bad individuals, good individuals etc. Variety should be there, only then one can understand the difference. If every thing is sought to be made uniform then the very essence of life is lost. Man should seriously think whether it is in the interest of nature and his own interest to pursue with unbridled genetic engineering research. Because, as per the process of natural selection, the genes that are not expressing in a given set of environmental conditions might express in a new environment and who knows what they might contain.